Strategies for Sustainable Technologies and Innovations

Strategies for Sustainable Technologies and Innovations

Edited by

John R. McIntyre

Georgia Institute of Technology, USA

Silvester Ivanaj

ICN Business School – CEREFIGE, France

Vera Ivanaj

Université de Lorraine – CEREFIGE, France

Edward Elgar

Cheltenham, UK • Northampton, MA, USA

Published by
Edward Elgar Publishing Limited
The Lypiatts
15 Lansdown Road
Cheltenham
Glos GL50 2JA
UK

Edward Elgar Publishing, Inc.
William Pratt House
9 Dewey Court
Northampton
Massachusetts 01060
USA

A catalogue record for this book
is available from the British Library

Library of Congress Control Number: 2012951764

This book is available electronically in the ElgarOnline.com
Business Subject Collection, E-ISBN 978 1 78100 683 2

ISBN 978 1 78100 682 5

Typeset by Servis Filmsetting Ltd, Stockport, Cheshire
Printed by MPG PRINTGROUP, UK

Contents

List of figures   vii
List of tables   viii
List of boxes   ix
List of contributors   x
Acknowledgements   xvii
Introduction   xix

PART I   BUILDING SUSTAINABLE TECHNOLOGY AND
         INNOVATION SYSTEMS

1. Sustainable innovation responses to global climate change   3
    Paul Shrivastava
2. Understanding eco-innovation for enabling a green industry
    transformation   21
    Tomoo Machiba
3. Sustainable development through innovation? A social
    challenge   51
    Corinne Gendron
4. Appraisal of corporate governance norms: evidence from
    Indian corporate enterprises   74
    Rabi Narayan Kar
5. Codes of conduct and other multilateral control systems for
    multinationals: has the time come – again?   100
    Tagi Sagafi-nejad
6. Appropriate technology movement   118
    Sanjeeb Kakoty

PART II STRATEGIC IMPLICATIONS AND
         ASSESSMENT

.7. Eco-social business in developing countries: the case for
     sustainable use of resources in unstable environments   139
     Roland Bardy and Maurizio Massaro

8. Entrepreneurship development at a small scale: a key to
   sustainable economic development   168
   Sanjay Bhāle and Sudeep Bhāle
9. Entrepreneur profile and sustainable innovation strategy   186
   Sandrine Berger-Douce and Christophe Schmitt
10. Benchmarking sustainable construction technology in the
    building and transportation sectors   204
    Salwa Beheiry and Ghassan Abu-Lebdeh
11. The eco-logistics improvement in France: towards a global
    consideration of inland waterway transport within the supply
    chain strategy   219
    Thierry Houé and Renato Guimaraes
12. Integrating sustainability and technology innovation in
    logistics management   239
    Matthias Klumpp, Sascha Bioly and Stephan Zelewski
13. Sustainable development, a new source of inspiration for
    marketing innovation? Focus on five major trends and one
    innovative project in customer relationship marketing   262
    Gaël Le Boulch and Rémy Oudghiri

Index   277

Figures

1.1 The interaction of rapid climate change and habitat fragmentation 7
1.2 Multinational corporations are systems of vision, inputs, throughputs and outputs 11
1.3 Industrial ecosystem at Kalundborg, Denmark 15
1.4 A schematic for thinking about sustainable innovation 17
2.1 The scope of Japan's eco-innovation concept 25
2.2 A proposed framework of eco-innovation 27
2.3 Conceptual relationships between sustainable production and eco-innovation 29
2.4 Diverse types of eco-innovation: incremental to systemic 31
2.5 Mapping primary focuses of eco-innovation examples 37
2.6 Various factors surrounding eco-innovation 41
2.7 Overview of policy instruments for eco-innovation 45
3.1 A new representation of sustainable development 55
3.2 An illustration of the seven core subjects of ISO 26000 standard 61
9.1 Creator profile and sustainable innovation strategy 192
11.1 The three dimensions of a sustainable logistics 221
11.2 The evolution of IWT in France from 1999 to 2009 225
11.3 Types of goods carried by IWT in France 226
12.1 Logistics market volume and functional segments 240
12.2 Sustainable investments in logistics 242
12.3 Stock keeping centralization (Germany) 243
12.4 CO_2 emissions caused by the electricity mix in Germany 254
12.5 Technology Innovation and Sustainability Assessment in Logistics (TISAL) 257

Tables

2.1   Industry practices examined through the eco-innovation framework   32

2.2   Application of technologies in different types of innovation   38

2.3   Business model eco-innovation: change in elements   43

4.1   Nature of investigations taken up by SEBI   88

4.2   Nature of investigations completed by SEBI   89

9.1   Comparative study of different treatment processes for hospital waste   195

9.2   ECODAS's key figures since 2004   196

10.1   Technology use   215

11.1   Most important specification of different costs according to transport modes   232

12.1   Results of insourcing vs outsourcing   245

12.2   Anthropogenic greenhouse gases emitted   251

12.3   Types of power plants in CO_2 comparison   255

Boxes

2.1   *Vélib'*: self-service bicycle-sharing system in Paris   34
2.2   The development of advanced high-strength steel for automobiles   35
2.3   Energy-saving controller for air conditioning water pumps   36
2.4   Examples of 'green business models'   40

Contributors

Ghassan Abu-Lebdeh is an Associate Professor of Civil Engineering at the American University of Sharjah (AUS) in the UAE, and former recipient of the Eisenhower Doctoral Fellowship at the University of Illinois at Urbana-Champaign, USA, where he obtained his PhD in Civil Engineering/Transportation in 1999. He earned a Master of Science degree in Civil Engineering from the University of New Brunswick, Canada, and a Bachelor of Civil Engineering degree from Yarmouk University in Jordan. Dr Abu-Lebdeh's research interests are in control and modeling of interrupted flow and congested systems, and advancement of sustainability in transport systems through advanced computing and modeling. Prior to joining AUS, Dr Abu-Lebdeh was on the faculty of civil engineering at the University of Kentucky, USA from 1999 to 2001, and at Michigan State University, USA from 2001 to 2008. He has ten years of industry experience in transportation planning and engineering at both the Central Massachusetts Regional Planning Commission in Worcester, MA, USA and the Champaign County Regional Planning Commission in Urbana, IL, USA.

Roland Bardy is owner of BardyConsult in Mannheim, Germany, where he mainly engages in management education, and he serves as Executive Professor of General Management and Leadership at Florida Gulf Coast University, USA. Born in Vienna, Austria, he received his MBA degree there in 1969, and his PhD degree (in econometrics) from Heidelberg University, Germany, in 1974. He worked in finance and administration at BASF SE, the German multinational chemicals manufacturer, for about 30 years until 1999. His areas of interest are management accounting, supply chain management, leadership and business ethics. Residing in both Mannheim, Germany, and Naples, FL, USA, Dr Bardy is privileged to experience both US and European developments.

Salwa Beheiry is Assistant Professor of Civil Engineering at the American University of Sharjah, UAE. She obtained her PhD in Civil Engineering from the University of Texas at Austin, USA, in 2005. She is a recipient of various honors and awards throughout her academic and industrial career. Before starting her Doctoral program, she worked with

Independent Project Analysis Inc. in Ashburn, VA, USA for five years as a project analyst/consultant in international building, industrial and infrastructure projects. Dr Beheiry earned a Master of Science degree in Project Management from George Washington University, USA in 1998 and a First Class Honors Bachelor of Science Degree in Construction Engineering and Management from the University of Reading, UK in 1994.

Sandrine Berger-Douce is a Professor of Management and Global Performance, Henri Fayol Department in National Institute of Science and Technology, Ecole des Mines in Saint-Etienne, France. She is a researcher in the SPICE (Sustainable Performance, Innovation and Change in Enterprises) team specialized in analysis and implementation of innovation processes, for a global performance of organizations to meet sustainable development and various stakeholders' interests. An author of academic articles, she is conducting several research projects on sustainable development and corporate social responsibility implementation.

Sanjay Bhāle is a full-time Director of Academics and Professor of Management with MIT School of Telecom & Management Studies, Pune, India and also a PhD guide at Symbiosis International University, Pune. He also holds an MBA and an MSc in Applied Chemistry. Before turning to academia Sanjay worked as Chief Chemist with renowned chemical companies. His teaching subjects are marketing management, entrepreneurship development, and product and brand management. Sanjay has published a number of research papers at the international level on subjects related to sustainability, entrepreneurship, innovation, supply chain and special economic zones. Sustainability and innovation are his research areas of interest.

Sudeep Bhāle is a full-time Project Manager and Project Office Manager working for one of the leading global engineering groups, based at the Sydney, Australia office. He has previously worked in sales and marketing and technical roles. Sudeep spent several years of early career working as a marine engineer, travelling around the world. Sudeep is a mechanical engineer, also holds an MBA and is a higher degree research student at Macquarie University Sydney, Australia. Sudeep's passion is to build a self-sustainable model of social entrepreneurship which could be customized to suit different economies of the world.

Sascha Bioly, after his high school degree (Abitur), served as a Telecommunications Officer in the German army. After that he worked as a systems administrator in information technology (IT). Trained as an industrial clerk, in 2005 he joined Dresdner Cetelem Bank in Duisburg,

Germany and worked as an Assistant, Deputy Team Leader and finally as head of the debt collection group. In 2006 he began his studies in business administration, completed in 2009. Subsequently he became a Research Assistant at the ild Institute of Logistics and Service Management, Germany.

Corinne Gendron is a full Professor at the University of Québec in Montreal (UQAM), Canada, and is the Social Responsibility and Sustainable Development Chair. Her courses include research methodology, firm conceptualizations and environmental and social management. She received her PhD in Sociology at UQAM, after an MBA in Marketing and Finance. She is also a lawyer, since 1990. Her research focuses on the new regulation dynamic in post-ecological societies, sustainable development, corporate social responsibility and new social economic movements. She recently published *Regulation Theory and Sustainable Development* (Routledge, 2012), and has won several prices for her research and expertise.

Renato Guimaraes has a Doctorate in systems engineering and a Master's degree from Santa Catarina Federal University in Brazil. He teaches internal flow logistics. He has several years of experience in production management and teaches at the division of the ICN Business School in Metz which specializes in operations management, supply chain management, logistics, purchasing and industrial marketing.

Thierry Houé is an Associate Professor at ICN Business School, France. He holds a PhD in Management Sciences from the Université de Lorraine, France, and teaches supply chain management, project management and e-commerce. He is the Head of the Academic Department of Supply Chain and Information Systems Management, and leader of the postgraduate common core course in supply chain management (Programme Grande Ecole). His research interests lie in the analysis of logistics strategies within the supply chain, particularly the geographical, relational and sustainable development aspects. In his work, he mainly uses qualitative methods including discourse analysis and case study analysis. Thierry Houé is also a member of the Management Research Centre of the Université de Lorraine (CEREFIGE), France.

Silvester Ivanaj has a multidisciplinary background combining academic scholarship, research, entrepreneurship and management experience. He is Associate Professor of Information Systems at ICN Business School, Nancy, France. Prior to joining the ICN Business School, he was a researcher engineer and then an environmental consultant. He was a visiting scholar in 2004 and 2009 at the College of Management, Georgia Institute of Technology, Atlanta, USA. Dr Silvester Ivanaj has over 17 years of experience in management education, information systems development and entrepreneurship. His research interests focus mainly on information systems, sustainability assessments methods and recently on virtual teams. Dr Silvester Ivanaj received his Master of Science degree from the Polytechnic University of Tirana, Albania and his PhD from the Institut National Polytechnique de la Lorraine of Nancy, France.

Vera Ivanaj is Associate Professor of Management Science in the Chemical Engineering School (ENSIC) of the Université de Lorraine, France. She received her PhD in Management Science from the Université de Lorraine. Her current research interests include strategic decision-making processes, sustainable development, logistics outsourcing, entrepreneurship and management education, coaching, team building and diversity. Dr Ivanaj is a member of AIMS (Association Internationale de Management Stratégique) and AGRH (Association Francophone de Gestion des Ressources Humaines), two of the most important francophone scientific conferences on strategic management and human resources management.

Sanjeeb Kakoty was born and lives in the charming hill city of Shillong, India. His quest to understand the past is confined not merely to knowing what happened, but also how and why things happened. This led to a postgraduate degree in History from North Eastern Hill University, Shillong followed by a PhD in the History of Technology. He believes that proper understanding of the past should give mankind the ability to engineer desirable change instead of merely trying to adapt to change. To that end, he has worked as a teacher, writer and a documentary film maker and is currently a faculty member of the Indian Institute of Management, Shillong, teaching sustainability and communications.

Rabi Narayan Kar is a postgraduate and holds a Master of Philosophy degree from the Department of Commerce, Delhi School of Economics, University of Delhi, India. He was awarded a PhD degree by the Department of Business Economics, University of Delhi, in Corporate Mergers and Acquisitions. He is also a Fellow of the Institute of Company Secretaries of India. He is an Associate Professor in the Department of Commerce, Shaheed Bhagat Singh College (Delhi University) and has academic associations in different capacities with the Institute of Company Secretaries of India, the Institute of Chartered Accountants of India, the Department of Commerce, Delhi School of Economics, Jamia Millia Islamia and several other reputed management institutes.

Matthias Klumpp is an Associate Professor in Logistics and Service Management at the FOM University of Applied Sciences in Essen, the largest private business school in Germany. Since 2009 he has been Director of the Institute of Logistics and Service Management (ild) and his research interests and publications belong to the fields of logistics, supply chain management (SCM) and education, especially trends, innovations, operations research and evaluation of logistics concepts as well as qualification and training schemes. He is a member of the LogistikRuhr scientific committee for the German national research excellence cluster in logistics.

Gaël Le Boulch is both an academic and a manager with strategic and operational experience. After three years in the Strategy Department of the future Suez Group and one year in the Strategy Department of EDF during its international expansion, Gaël spent four years in operational management in France, preparing small businesses in the Paris region for competition in the electricity market. He is now a consultant specializing in strategy, organization and management. With a keen interest in innovation and research, Gaël Le Boulch has been a teacher at Paris IX Dauphine University, France for five years and is a research associate at CREPA, the CNRS strategy and management research institute. He has a doctorate in management sciences from Paris Dauphine University, France.

Tomoo Machiba has been Senior Programme Officer for Knowledge Management at the International Renewable Energy Agency (IRENA) since February 2012. Prior to joining the IRENA, he worked for the Organisation for Economic Co-operation and Development (OECD) as a Senior Policy Analyst in the Directorate for Science, Technology and Industry (DSTI) between 2008 and 2012. He managed a flagship project on Green Growth & Eco-innovation and worked on the DSTI's contributions to the OECD Green Growth Strategy. Beginning his career as journalist, Tomoo has been involved in the field of corporate social responsibility (CSR), energy and resource efficiency, and sustainable consumption and production (SCP) over 15 years. He served as an associate for SustainAbility, a leading CSR consultancy based in London, UK, and worked at the Global Reporting Initiative (GRI) headquartered in Amsterdam, as a program manager for technical development of the GRI Sustainability Reporting Guidelines. Most recently, he was a senior consultant to the United Nations Environment Programme (UNEP)/ Wuppertal Institute Collaborating Centre on SCP in Germany, working for the UN Marrakech Process. Tomoo holds an MPhil in Development Studies from the Institute of Development Studies (IDS), University of Sussex, UK.

Maurizio Massaro started his career as a business consultant and teacher of postgraduate and undergraduate courses before receiving his PhD in

Business from Udine University, Italy. He is co-founder and former Chief Executive Officer (CEO) of three small consultancy companies in the north-east of Italy. Having taught university classes and master classes since 2001, he officially joined Udine University as Aggregate Professor in 2008. He was a Visiting Scholar at Florida Gulf Coast University, USA in 2010, and apart from his university activities he teaches accounting on postgraduate and undergraduate courses at private business schools and in public institutions.

John R. McIntyre is Professor of International Management and International Affairs with joint appointments in the Ernest Scheller Jr College of Business and the Sam Nunn School of International Affairs of the Georgia Institute of Technology, Atlanta, GA, USA. He is the founding Director of the Georgia Tech Center for International Business Education and Research (CIBER), a US national center of excellence. He received his graduate education at McGill, Strasbourg and Northeastern Universities, USA, obtaining his PhD at the University of Georgia. McIntyre has had work experience with multinational firms in the UK and Italy. He is an elected member of the Board of Advisors, World Trade Center, Atlanta, GA. He is a consultant to international companies focusing on trade and investment strategies.

Rémy Oudghiri is the Director of the Trends and Future Studies department at Ipsos (Paris). He has a sociology and marketing background (HEC, IEP, DEA Sociology). An expert in monitoring consumer trends and values at both local and global level, he is in charge of syndicated observatories in manifold topics (e.g. luxury, consumer values, well-being) and is particularly experienced in planning and reporting large-scale international studies, particularly in luxury goods, and in the area of consumer lifestyles and values. He also regularly conducts consumer segmentations (e.g. luxury, food and drink, consumer electronics, mobile telephony).

Tagi Sagafi-nejad is a Radcliffe Killam Distinguished Professor of International Business, Director of the Center for the Study of Western Hemispheric Trade, USA, and Editor, *International Trade Journal*. He was Founding Director of the PhD program in International Business Administration at Texas A&M International University, 2003–08. He is also Professor Emeritus of International Business and former Chair of the Department of Management & International Business at the Sellinger School of Business & Management at Loyola College in Maryland, USA. He holds an MA and a PhD from the University of Pennsylvania, USA, where he also taught Doctoral-level courses and conducted research on technology transfer at the Wharton School.

Christophe Schmitt is currently a Professor of Entrepreneurship and holds the Chair of Entrepreneurship at the Université de Lorraine, France. He works in the field of student entrepreneurial culture development. His work focuses on the entrepreneurial process and entrepreneurial cognition. He is the author of many academic articles in this field. He has developed tools and methods such as IDéO in order to evaluate the potential of business opportunities. He has also written several articles and published books about entrepreneurship. His latest book deals with the notion of value of products and services in small and medium-sized enterprises (SMEs). He is also Associate Professor of CRIPMEE in Canada and Holy Spirit University of Kaslik in Lebanon.

Paul Shrivastava received his PhD from the University of Pittsburgh, PA, USA in 1981. He has a Post Graduate Diploma in Management at the Indian Institute of Management, Calcutta (1976), and a BE (Mechanical) from Bhopal University, Bhopal, India (1973). He is currently the David O'Brien Distinguished Professor and Director of the David O'Brien Centre for Sustainable Enterprise at the John Molson School of Business, Concordia University, Montreal, Canada. He also serves as Senior Advisor at Bucknell University and the IIM-Shillong, India, and leads the International Chair for Arts and Sustainable Enterprise at ICN Business School, Nancy, France.

Stephan Zelewski is a full Professor of Business Administration at the University of Duisburg-Essen, Campus Essen, Germany. He is Director of the Institute for Production and Industrial Information Management and has led a multitude of research projects, especially in the fields of artificial intelligence in production and logistics as well as knowledge management and motivation. He received his PhD and Habilitation from the University of Cologne in Germany, had a first professorship at the University of Leipzig, Germany and has published extensively in the fields of knowledge, motivation and game theory approaches to new production and logistics concepts.

Acknowledgements

The organizers of the research conference, and the editors and authors of this research volume, have amassed some not insignificant debts of gratitude to a number of individuals, colleagues and organizations whom they wish to acknowledge. This volume is mainly a direct offshoot of the November 4, 5 and 6, 2009 international conference titled 'Multinational Enterprise and Sustainable Development: Strategies for Sustainable Technologies and Innovation' (MESD 2009), held at the Palais des Congrès de Nancy, France. We wish to acknowledge the support of The Coca-Cola Entreprise, France; the PRME (Principles for Responsible Management Education); the ICN Business School and its administration; the Centre Européen de Recherche en Economie Financière et Gestion des Entreprises (CEREFIGE), of Université de Lorraine, France which mobilized its extensive regional research network; and the CIBER Centers program through its Georgia Tech Center for International Business Education and Research. We express our grateful thanks to the United Nations PRME, Atlanta, Xerox France, Schneider Electric, Rhodia, La Fibre Verte, abcvert, Inderscience publishers, ENSIC, MGIMO, Georgia Tech-Lorraine, the Lorraine Region, Communauté Grand Nancy, the Conseil Général de Meurthe et Moselle and finally to Metz Métropôle for their guidance and support in organizing the original event.

All of the authors included in this volume have worked long and hard in preparing their chapters and 26 reviewers were involved in two rounds of chapter selection and revision. We wish to single out keynote speaker Paul Shrivastava, David O'Brien Distinguished Professor of Sustainable Enterprise, and Director, David O'Brien Center for Sustainable Enterprise, John Molson School of Business, Concordia University; Mr Tomoo Machiba, past Senior Policy Analyst (Project Manager, Sustainable Manufacturing & Eco-innovation), Structural Policy Division (SPD) – OECD Directorate for Science, Technology and Industry (DSTI); Prof. John C. Crittenden, Director of Georgia Tech's Brook Byers Institute for Sustainable Systems, Georgia Institute of Technology; Prof. Corinne Gendron, Chairholder and Professor, Chaire de responsabilité sociale et de développement durable, ESG-UQÀM, Montreal, Canada; Prof. Tagi Sagafi-nejad, The Radcliffe Killam Distinguished

Professor of International Business, Director, Center for the Study of Western Hemispheric Trade and Director, International Trade Institute. Many thanks go also to Michael Matlosz, Director of ENSIC-INPL; Patrick Mazeau, Manager Customer Led Innovation, Xerox Innovation Group; Christian Wiest, International Operating Division, Executive Vice President, Schneider Electric; and Arnaud Rolland, head of the Coca-Cola Enterprises' sustainable unit, who were most helpful in their personal support and guidance.

A special word of thanks to Georgia Tech-Lorraine, the European campus extension of the Georgia Institute of Technology, in Metz, France which organized a reception for the participants of the original November 2009 event.

Introduction: Foundational considerations in balancing innovatory processes and sustainable development practices in comparative light

John R. McIntyre, Silvester Ivanaj and Vera Ivanaj

This book of expert and scholarly contributions on strategies for sustainable technologies and innovations is the by-product of the three-day international conference (Multinational Enterprises and Sustainable Development – MESD'09) held at Nancy, France, in November 2009, jointly organized by the ICN Business School (France), the CEREFIGE research center of the Université de Lorraine (France), and the Georgia Tech Center for International Business Education and Research, Georgia Institute of Technology, Atlanta, USA.

The conference brought together academics, research-oriented practitioners, experts, consultants and various professionals in the field of technology management for sustainable development with a view to refine our understanding of one of the major challenges of our environmental future: its innovation dimensions. Our conviction is that the form in which business operations are actually conducted around the world cannot be sustained, as momentous changes continue to characterize our planet. Scientific evidence points to the fact that human and organizational behaviors can result in critical damage to our natural systems. The quality of human existence continues to be strongly related to our ability to innovate and to conceive sustainable technological and productive systems. Radical socio-technical changes are needed to slow and eventually reverse the deterioration of our environment, but also to develop available natural resources. Business firms are catalysts and agents of societal and economic change. Over the last decade, companies have faced social and environmental pressures to better integrate the challenges of sustainability. Scholars, policy-makers and experts, among others, have argued that sustainable development is the perfect opportunity for businesses to strengthen the evolving notion of corporate social responsibility, while achieving long-term growth through the innovatory process and capitalizing on research and development. Companies that ignore these opportunities do so at their peril. But companies have had considerable difficulties in addressing the opportunities inherent in the challenge of sustainable development. In particular, their technological and innovation strategies are often inappropriate to manage the complex and the uncertain nature of this new demand. Corporate strategies that integrate the goals of sustainability are urgently needed.

Given the ever-changing global economic environment and the challenges raised by the legal process, strategic choices for technology and innovation have become key factors for success in implementing sustainable development policies. The economic growth and competitiveness as well as the societal well-being of businesses can be predicated on these choices and policies.

This book focuses on the issue of strategies for innovation-driven sustainable technologies. It seeks to address the following major questions regarding sustainable development: Why, when, and how will such strategic technological and innovatory choices be made and deployed? Can enterprises make technological choices that will be economically advantageous, ecologically sustainable and socially responsible? How do corporations balance and harmonize their choices, considering the elements of technological innovation, economic growth, resource efficiency and environmental protection? Do contextual, economic, ecological or societal factors play a role in economic and financial profitability, competitiveness, market openness, policies, technical standards and regulations? Are there international differences that bear noting and have some explanatory power? Are current decisions an adequate response to anticipated future needs? What is the relationship between technological strategies employed by businesses and a country's or a sector's economic performance? How do enterprises employ strategic analysis in order to compare the negative impacts of present-day technologies with the positive benefits of future innovations?

These questions are at the core of sustainable strategic management and apply to management scholars, economists, lawyers, sociologists and individuals in all the relevant engineering fields. It must be reiterated that while multinational corporations have been at the forefront of sustainable development, the role of small and medium-sized enterprises cannot be undermined as they respond to the broad trends set in motion by larger firms. Nor can the ethical dimension of such choices, strategies and modes of implementation be ignored as they frame the debate. Technology sets the production frontier, and should be viewed as a neutral factor, often resulting from the political stakes of public and private decision-makers.

Our book seeks to provide guidance to enterprises of all sizes and to organizational decision-makers seeking sustainable technology and innovation-driven solutions. The purpose of the authors and of the entire volume is to explore and share ways in which business firms can bring their technology and innovation strategies in closer alignment with the requirement of sustainability.

The book is structured in two parts, each of which deals with a specific aspect of sustainable technology and innovation systems: 'Part I: Building sustainable technology and innovation systems', and 'Part II: Strategic implications and assessment'.

PART I: BUILDING SUSTAINABLE TECHNOLOGY AND INNOVATION SYSTEMS

The six chapters in Part I address the relationship between sustainable technology and innovation in the context of global socio-economic systems. The core idea is that sustainable technologies and innovations must benefit both companies and global economic and social development. Innovation can be a value-creating process and can integrate social needs more fully through a radical, or at least momentous, cultural change by empowering both customers and employees. This process of innovation has evolved a transformation in which multinational corporations of necessity will play a central role responsive to a more globalized regulatory system. Moreover, life philosophy, beliefs and values of the world community will also shape sustainable technologies and innovation systems.

In Chapter 1, Paul Shrivastava examines how sustainable innovation constitutes a useful response to global climate change. Climate change is rooted in human activities, many of which are directly controlled by corporations. Sustainable innovations can assist companies in improving their ecological and social performance and gaining competitive advantage. He shows that a mindful application of appropriate technologies and rational use of resource conservation by companies can yield large gains in performance. This, he claims, can happen with respect to a company's vision, inputs, throughputs and outputs. Past literature associates technological innovations with 'high-tech', and large research and development (R&D) budgets. Shrivastava asserts that this is not necessarily the case. Sustainable innovations can take simple rationalization approaches, de-bureaucratizing organizational procedures and conserving renewable resources sensibly. He provides several examples of sustainable innovations in main-line companies such as General Electric, Walmart, 3M Corp, United Parcel Service of America Inc., and others. Sustainable innovations can also be facilitated by empowering customers and employees to eliminate wasteful practices in areas where small benefits can be magnified by frequency of use.

In Chapter 2, 'Understanding eco-innovation for enabling a green industry transformation', Tomoo Machiba elaborates a theoretical framework of eco-innovation by taking stock of the extant knowledge, and proposes a fresh perspective for more comprehensive analyses in the future. The chapter first outlines a context of green growth and eco-innovation in the past two decades since sustainable development emerged as a global policy discourse. Green growth can be considered as a new strategy to revive the sustainability agenda and to integrate it practically into mainstream policies as well as industry activities.

Green growth generally entails decoupling economic growth from environmental degradation. However, in many areas, environmental pressures continue to rise as economies grow, and improvements in efficiency have often been offset by increasing consumption and outsourcing. The challenges cannot be met by business as usual. In this chapter, eco-innovation (or sustainable innovation) is given as a key factor to enable the decoupling. Diverse types of eco-innovation exist including both the creation of new technologies, products and processes, as well as their application and diffusion. Synthesizing the existing understanding and definitions, the author proposes to analyze eco-innovation based on a three-dimensional framework that consists of an innovation's target, mechanism and impact.

An examination of corporate sustainable production activities through this proposed framework indicates that the primary focus of current eco-innovation tends to rely on technological advances, typically with products or processes as eco-innovation targets and with modification or redesign as principal mechanisms. Nevertheless, a number of complementary organizational or institutional changes have functioned as key drivers for these developments, such as the setting up of intersectoral or multi-stakeholder collaborative networks. The author argues that a sophisticated combination of different types of innovation could bring far-reaching changes in the techno-social system and enable a long-term green transformation by impacting several components of the economy including consumers.

Among the key social, technical and political elements in determining the success of eco-innovation, particular attention is paid to the 'business model', which drives eco-innovation to the market and enables its diffusion. The business model perspective allows a deeper and more subtle understanding of how environmental value is captured, turned into profitable products and services, and can deliver convenience and satisfaction to users. It is particularly relevant to radical and systemic eco-innovation as it also makes for a sharper understanding of eco-innovation as one of the paths to restructure the value chain, producer–consumer relationships and evolving consumer practices.

Lastly, the role of diverse policy instruments in the acceleration of eco-innovation is reviewed. Many radical and systemic solutions have been facing rather high entry barriers since they do not always fit the existing technology systems and need long-term investment in the development of new infrastructures as well as cultural changes. To break the technological lock-in and path-dependencies, policy decisions made today need to incorporate a longer time horizon and governments need to take a stronger leadership position in actively supporting infrastructure and platform development. Exemplars of these include the development of smart grids, public transport systems and green cities.

In Chapter 3, Corinne Gendron argues that sustainable development has been discussed for decades now, but its relationship to the national economy still remains unclear, and even controversial. With the new concept of the 'green economy', the United Nations Environment Programme (UNEP) is proposing an insightful clarification about the role it must play to reorient our societies towards sustainable development. Building on the Porter hypothesis, the UNEP report supports a green and inclusive growth enabled and potentiated by innovation. But to play such a central role, technology has to be driven by societal needs through a process of transition management. This requires deepening our understanding of the relationship between innovation and social and cultural changes which may constitute more powerful drives than traditional technology.

In Chapter 4, 'Appraisal of corporate governance norms: evidence from Indian corporate enterprises', Rabi Narayan Kar reviews how the issues relating to corporate governance have come into prominence because of their evident importance to the sustainable health of corporations as well as society in general, in the context of the plethora of corporate scams and debacles in the recent times. The United States (US), Canada, the United Kingdom (UK), European countries, East Asian countries and India have all experienced several pressures on their economic well-being and have faced the ever-widening impact of the Great Recession of 2008 in its global dimensions, leading to the demise of several of their leading companies. Already a greater emphasis on corporate governance issues has been set in motion. Corporate governance broadly refers to a set of strategies and best practices designed to govern the behavior of corporate enterprises and light their path in a forecasting perspective. Corporate governance deals with laws, procedures, practices and implicit rules determining a company's capacity to make managerial decisions endowed with a long-term perspective and to implement innovative strategies oriented to sustainable growth. In this context, governments all over the world are playing the role of facilitators by supporting, and often promoting, good corporate governance as well as the ethical functioning of corporate enterprises. This chapter analyzes the regulatory framework of the corporate governance system in India. The case study of Satyam Computers has been incorporated to sharpen our grasp of the issues and to draw out insightful comparisons with other corporate failures.

Chapter 5 authored by Tagi Sagafi-nejad tracks the evolving relationship between multinational enterprises (MNEs) or transnational corporations (TNCs) and governments, from the 1970s when a protracted exercise under United Nations (UN) auspices to establish a code of conduct for MNEs was ultimately abandoned, and through the 1980s when an OECD initiative to establish a multilateral agreement on investments was likewise given up. At the dawn of the twenty-first century, and in the wake of the 2008 global financial crisis, the need for multilateral instruments to delineate boundaries and establish rules of engagement for MNEs and nation-states re-emerged with momentous salience. What can be learned from earlier attempts to establish such rules? Has the time come for promulgating some form of international accord concerning the entire gamut of nation-states' relations with these enterprises? The chapter draws parallels between the turbulent 1970s and the current uneasy relations between nation-states and TNCs, and explores the extent to which lessons learned during the earlier period can lead to a better understanding of the more complex and nuanced international business environment of the twenty-first century. The chapter asserts that earlier attempts to establish such rules are fraught with actionable lessons, and concludes that the time has come to evolve some form of international accord concerning the entire gamut of nation-states' relations with multinational border-crossing enterprises.

Sanjeeb Kakoty in Chapter 6, 'Appropriate technology movement', examines how history and human experience highlight the preponderant influence of leading technologies in succeeding periods of time. Interestingly, the essential fabric of human life and social mores was often built around the predominant technology extant during that time. But the question that arises is: What determines the choice of technology? Does technology arise due to the specific physical needs of the community? Is technology in turn shaped and influenced by the philosophy, religious beliefs and worldview of a particular leading community? In this scenario, what kind of impacts would imported technology have on the specific need mitigation of the community, and also on their philosophy of life? Would it be possible and feasible to have a uniform technology for all regions of the world, or would it be more suitable for disparate regions and individual communities to develop their own unique region-specific technologies? This is the crux of the appropriate technology movement. Schumacher is considered to have exerted the preponderant influence on the movement by formulating compelling arguments for small-scale, decentralized, environmentally sustainable enterprises. He, in turn, was greatly influenced by the votary of truth and non-violence Gandhiji, who felt that economics and ethics were two sides of the same coin. These thoughts are also a reflection of Hindu and Buddhist traditions.

By the dawn of the new millennium, serious doubts had emerged about the rationale behind man seeking perpetual economic growth through rapid industrialization, often ignoring the human and environmental costs. What resulted was a search for an alternative blueprint for progress. This model had to be based on the use of appropriate technology which involved taking cognizance of local needs, conditions and aspirations. Interestingly, these considerations automatically brought with them the principles of sustainability and best business practices.

PART II: STRATEGIC IMPLICATIONS AND ASSESSMENT

The seven chapters contained in Part II deal with the contextual factors defining how to implement sustainable technologies innovation. Macroeconomic factors such as the economic level of the country development, or microeconomic ones such as business size and the industry organization, are taken into account. Part II of the book focuses on how effective innovation and technology strategies can be crafted in developing countries, and also for smaller businesses or entrepreneurs. Innovative business opportunities exist not only for core organizational functions such as product innovation, but also for different support functions, such as the supply chain and logistics.

Chapter 7, 'Eco-social business in developing countries: the case for sustainable use of resources in unstable environments', authored by Roland Bardy and Maurizio Massaro, explores how the concept of eco-social business, generating productive employment opportunities while supporting a responsible use of natural and non-renewable resources in developed countries, can be applied to emerging and transition economies; and how this approach can be buttressed through foreign direct investment (FDI) or other forms of border-crossing business interventions. The authors take up the question whether a realistic and effective business strategy addressing the issue of sustainable use of resources in the developing world is possible. They argue that business opportunities can be found for most investors in developing economies, either through creating new markets with appropriate innovative products or services, or through new forms of production and by building local capabilities utilizing local natural resources. The chapter further demonstrates which research domains must be brought together, offering a conceptual framework for eco-social business in developing economies.

Chapter 8, 'Entrepreneurship development at a small scale: a key to sustainable economic development', authored by Sanjay Bhāle and Sudeep Bhāle, looks into the power of collective efforts made at a small scale and how they can be converted and scaled to a large, sustainable, growing business supportive of grass-roots initiatives and actors. A case study is also presented, 'A modest testimony', on how an organization started by seven women with a sum of less than US$2 grew over a period of years into a multi-million-dollar organization. This example is illustrative of the paradoxical notion that 'small is big'. The chapter also advances some basic fundamentals of social entrepreneurship (SE) and how social entrepreneurship can contribute to sustainable development of societies. The 'key features of social entrepreneurship' embedded in this chapter provide some productive insights to this new emerging field of entrepreneurship, capturing salient attributes related to social entrepreneurship, and offering examples of great work in this area. Overall this chapter is a demonstration of the varied possibilities evident in different economies to build a sustainable economic environment by bringing together the right mix of ideas, effort and entrepreneurial attributes.

In Chapter 9, 'Entrepreneur profile and sustainable innovation strategy', Sandrine Berger-Douce and Christophe Schmitt describe a specific project illustrating a sustainable innovation strategy. That issue concerns the study of social responsibility and innovation in small and medium-sized enterprises. Among the multiple issues involved in sustainable strategies, Chapter 9 focuses on the reasons why economic representatives choose to carry out these strategies. In small and medium-sized enterprises, the literature shows that the entrepreneur's profile is a dominant influence on strategic choices. This chapter analyzes this relationship in some depth. A case study was conducted at Ecodas, a small business located in northern France. Created in 2000, the firm is specialized in the design and manufacturing of medical waste treatment machines using shredding and sterilization technologies. The case of Ecodas illustrates how to reconcile economic performance with ecological respect and social responsibility, in accordance with the international norm ISO 26000 published at the end of 2010. The authors report that the key factors for a sustainable innovation strategy for small businesses are based on elements of the entrepreneur's profile (social capital, audacity and tenacity) supported by a favorable regulatory environment. In other words, this chapter underlines the necessity to combine an entrepreneur's profile with a favorable environment in order to succeed with the sustainable technology-oriented innovations.

Chapter 10, 'Benchmarking sustainable construction technology in the building and transportation sectors', authored by Salwa Beheiry and Ghassan Abu-Lebdeh, deals with the principles of benchmarking sustainable construction technologies and techniques in the building and transportation sectors. This industry sector targets residential and commercial property and transportation projects, including roadway, bikeway and walkway segments; junctions including roundabouts, bridges including piers, signalized and non-signalized intersections, tunnels, runways, taxiways and holding areas; and facilities including bus stops and rail and metro stations, and transfer points and pavements. The chapter also discusses process benchmarking as a vital tool in measuring the performance of capital projects and, in recent years, also in measuring the advancement of sustainable technologies and tools in building and transportation projects. Benchmarking provides organizations with a means to gather information about the current levels of application of sustainability practices and technology integration in project execution. Furthermore, the chapter gives insight into the evolving progress of the sustainability movement in the different engineering sectors. Chapter 10 also discusses a sustainable technology benchmarking case study, using a pilot data sample and a follow-up data sample of residential and commercial buildings in the United Arab Emirates. It is this type of industry-specific study that can strengthen and empirically confirm the theoretical constructs articulated in this text.

Chapter 11, authored by Thierry Houé and Renato Guimaraes, argues that key elements of the optimization of supply chains and transportation policy are necessarily linked with the sustainable development strategy of companies. This chapter considers the integration of inland waterway transport within the supply chain organization implemented by sustainable firms. Through a literature review and some practical examples based in the French context, it tries to demonstrate how the use of waterways, especially in combined transport operations, leads to an improvement of sustainable and innovative practices for supply chain management. However, even if eco-logistics processes are often presented as a considerable asset, their complexity is likely to generate significant and, at times, prohibitive costs and risks. These pages also look at the need for companies to consider new sustainable solutions to manage their flows, taking into account the complexity of today's international logistics and the nature of the product transported. On the one hand, the chapter concludes that the key factors for success of this integration are related to improved relationships between the different players in the global supply chain (shippers, logistics services providers, and so on) and to reorganizing models of transport. On the other hand, it shows that the use of inland waterway transport cannot be generalized without a strong political will but also an expansion of dedicated and sustainable multi-modal services.

In Chapter 12, 'Integrating sustainability and technology innovation in logistics management', Matthias Klumpp, Sascha Bioly and Stephan Zelewski discuss the prima facie conflict between technological innovation and sustainability in logistics. The authors provide an introduction to the logistics market and growth of the value-added service industry of logistics and supply chain management. They also show that in specific situations, warehouse centralization sustainability is hard to achieve today, based on existing information and cost structures. They consider the example of enhanced transparency in the supply chain, for example using innovative radio-frequency identification (RFID) technology to build sustainable logistics solutions. This research contribution urges logistics managers involved in long-term plans and projects to integrate a sustainability perspective into their investment calculations as sustainability on an operational level is largely dominated by long-term investments, that is, warehouse and plant locations or transport network structures.

Finally, Chapter 13, 'Sustainable development, a new source of inspiration for marketing innovation?' by Gaël Le Boulch and Rémy Oudghiri argues that while 46 percent of the French feel that their companies do not do anything for the environment, politicians increasingly seem to take sustainable development issues into account. In the US under the leadership of President Barack Obama, and in Europe where green movements have made a significant breakthrough in the recent European elections, this issue is high on the policy agenda, more than a mere fad. Nevertheless, companies – whatever their size – seem to have some difficulty in revising their business models and assumptions. How are companies to integrate sustainable development in the business processes and not to reduce it to a pure marketing tool? The truth is that customers and consumers often perceive an egregious lack of commitment. This chapter aims to present five ways to implement sustainable development in companies. How are companies to assure that sustainable development impacts directly on the corporate bottom line and is an organic part of their innovation strategy? This chapter finally considers how a major public company can reinvent itself and its client relations through innovative sustainable development decisions.

CONCLUSIONS

This book has the merit of addressing how researchers and practitioners think through the complex nexus of sustainable technology and innovation questions, from theoretical, regional, corporate and industry 'best practices' perspectives.

Selected contributors have argued that we can make technology and innovation choices that are profitable for firms and promote economic development for countries in development while balancing the environmental protection and social progress equities. Companies in developed and emerging countries are conceiving and implementing sustainable strategies, taking into account contextual factors to find appropriate ways of doing business that do not harm the Earth's natural equilibrium. Profitability oriented strategies as well as ecologically conscious strategies, as is argued and demonstrated, can converge toward a more holistic process of doing business. As a consequence, research and empirical methodologies which describe and explain the range of multidimensional influences are essential building blocks to achieve balanced sustainability strategies supported by the innovatory process.

PART I

Building sustainable technology and
innovation systems

1. Sustainable innovation responses to global climate change

Paul Shrivastava

INTRODUCTION

Sustainable development is a complex idea. It refers to managing our organizations or economies and ourselves in such a way that we can meet our own needs without jeopardizing the ability of future generations to meet theirs (Brundtland, 1987). Corporations and particularly multinational corporations will need to play a central role in achieving sustainability as they are currently the main engines of economic development and growth (Shrivastava, 1995).

Enterprises have the financial resources and the technological knowledge to innovate in ways that make them more sustainable. Innovation literature emphasizes the role of research and development (R&D), new high-tech investments and large-scale technological changes (Haanaes et al., 2011; Nidumolu et al., 2009; Nelson and Winter, 1982). Technology has been a key element of innovation thinking for the past several decades. Sustainability involves changes of both production and consumption. Therefore, a view of sustainable innovations needs to be broader than just technological. It needs to include social, cultural and political practices that reduce the eco-footprint of companies and people (Ehrenfeld, 2009).

Sustainable innovation (SI) does not always need to be high-tech and high-investment options requiring large amounts of funding in R&D. Innovation comes from rethinking ways in which things are currently done and by improving them incrementally. Our approach to SI seeks sustainability goals by mindfully implementing technological and social efficiencies. This means improving the functional, ecological and social performance of products, services and processes. There are many avenues for such improvements, including eliminating duplications, redundancies and waste; changing raw materials, equipment and packaging procedures; redesigning systems, separating or consolidating functions; and debureaucratization.

Sustainable innovations are often rooted in simple changes, which can be magnified to address large complex problems. To focus my discussion on this powerful possibility, I will explore SI in the context of one of the most complex sustainability challenges of our time: global climate change. By applying awareness and mindfulness, significant improvement can be made in sustainability and climate change mitigation strategies.

Sustainability is a very broad topic involving population explosion, environmental pollution, natural resources, species extinction and energy use. Therefore, it would be helpful to narrow down this discussion to a specific and important challenge of sustainability: global climate change. Global climate change is at the heart of sustainable development and if methods are not developed to deal with climate change challenges, there is no way to achieve sustainable development.

This chapter will begin by examining the threat of global climate change by providing some background scientific information along with information on the predicted and observed impacts on varying biomes. Some of the main international conferences and conventions addressing climate change are then described. Afterwards, I argue that sustainability is a crucial solution to the global climate crisis, and that many of the underlying concepts of sustainability apply directly to climate change mitigation procedures. In addition, some of the main challenges of sustainability are addressed. This is followed by a discussion on sustainable innovation and some examples of where and when it has been applied. Finally, I conclude the chapter with some suggestions for companies to implement innovation, and propose key mechanisms to encourage and involve employees in the sustainable innovation process.

GLOBAL CLIMATE CRISIS

There is sufficient evidence and consensus among the scientific community and the general public that the global climate is changing, caused by human-induced greenhouse gas emissions. There are several statements issued by global scientific panels confirming the threats of climate change. These include: (1) Joint Statement of Science Academies of G8 plus Brazil, Russia, India and China – Global Response to Climate Change, 2005 (Science Academies, 2005); (2) The Intergovernmental Panel on Climate Change, *Climate Change 2007: Synthesis Report* (IPCC, 2007); and (3) Sir Nicholas Stern's Stern Review on the Economics of Climate Change (Stern, 2007). The same agreement has been reached by the World Business Council for Sustainable Development, an organization of 800 chief executive officers (CEOs) of major companies; and by the National Aeronautics and Space Administration (NASA) in the United States, which confirms the data measuring physical changes caused by climate change.

The scientific facts can be briefly summarized as follows. Atmospheric carbon dioxide (CO_2) levels for the past 650 000 years have remained within a range of 180 ppm to 300 ppm due to natural variations in the climate system. This range, however, began to change with the onset of the industrial revolution in the mid-1700s. Concentration levels increased from 280 ppm in 1750 to 380 ppm today. Atmospheric CO_2 naturally traps heat in the atmosphere and warms the planet, also known as the greenhouse effect. However, recent high levels of CO_2 have enhanced this process and caused global surface air temperatures to rise. Historical records indicate that during the twentieth century the Earth warmed by 0.6°C. Additional warming is also expected in the twenty-first century, as much as 1.4 to 5.8°C above the 1990 level. This global warming causes melting of glaciers and ice sheets, rise in sea level, changes in crop productivity, and changes in occurrence and transmission of viruses and disease vectors. The Intergovernmental Panel on Climate Change (IPCC) Fourth Assessment Report (FAR) states that the upper limit of safe greenhouse gas concentration is somewhere between 300 and 350 ppm (IPCC, 2007).

The IPCC Climate Change Synthesis Report (IPCC, 2007) reflects global temperature spanning 200 years according to eight different climate change research centre models. The forecasted warming will increase the rate of ice melting and enhance the thermal expansion of seawater, leading to an increase in global sea levels. The IPCC has estimated that the global sea level will rise by 0.1–0.9 meters by 2100, relative to the level in 1990. A 0.5 metre rise in sea level in Bangladesh would put approximately 6 million people at risk of floods and as many as 200 million people would be displaced by droughts, sea level rise and floods by 2080 (IPCC, 2007). The same report demonstrates the sea-ice melt currently being experienced in Greenland and Antarctica.

PHYSICAL EVIDENCE OF GLOBAL CLIMATE CHANGES

Evidence of global climate change has been confirmed not only by empirical data and historical modelling but also by present-day extreme weather events. For example, the World Meteorological Organization (WMO) in 2010 stated that 11 of the previous 12 years had been the warmest since 1850; 2005 being the hottest year on record, followed by 2007 and 1998.

Climate change is affecting global hydrological systems, where average ocean, river and lake temperatures have increased. Also, climate change strongly influences the marine coastal environment and acts in synergism with other anthropogenic pressures to alter the state and functioning of biological and ecological systems (Goberville et al., 2010). Since oceans absorb nearly 80 per cent of the heat added to the atmosphere, their role in the climate system is critical. Other hydrological impacts such as increased runoff and earlier spring peak discharge in snow-fed rivers have compromised water quality (Cramer et al., 2001).

Marine biological impacts have also been reported due to global climate change. For example, there have been observed increases in algae and plankton, and reduced marine biodiversity in high-latitude oceans and lakes (Fox et al., 2009). Earlier migration shifts have also been documented in rivers. These observed changes in marine and freshwater biological systems are associated with rising water temperatures, as well as related changes in ice cover, salinity, oxygen levels and circulation (Fox et al., 2009).

Seasonal shifts have impacted upon terrestrial biological systems, where early springs have triggered leaf-unfolding, bird migration and egg-laying. The observed early 'greening' of vegetation in the spring is linked to longer thermal growing seasons and increasing temperatures, creating pole-ward shifts in biological ranges (Morin et al., 2009). At the same time, increased rates of temperature change have surpassed the adaptive capacity of many species and caused intense selection pressures and overall reduced genetic diversity (Jump and Penuelas, 2005). Current problems of habitat fragmentation coupled with rapid climate change have resulted in decreased growth and survival rates, and ultimately increased extinction risk (Jump and Penuelas, 2005; and see Figure 1.1).

Changes have been recorded in some Arctic and Antarctic ecosystems, including those in sea-ice biomes and among predators at high levels of the food web. There have been an increasing number and enlargement of glacial lakes, ground instability in permafrost zones and rock avalanches in mountain regions (Weller, 1998).

Some of the projections for global climate change include global warming of 0.3°C per decade and warming over the twenty-first century of 2.0–5.4°C (IPCC, 2007). Additional warming will only exacerbate the above-mentioned environmental problems.

SUSTAINABILITY: A RESPONSE TO GLOBAL CLIMATE CHANGE

Environmental destruction has been under discussion globally since the 1970s. Since then there have been several international conferences and

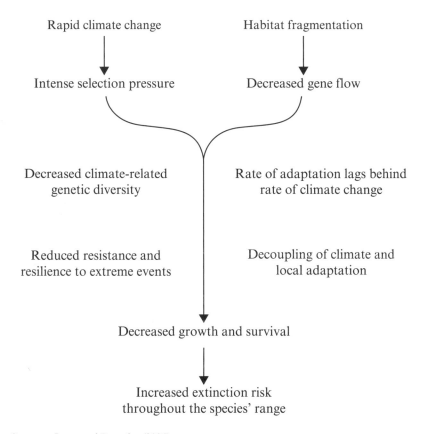

Source: Jump and Penuelas (2005)

Figure 1.1   The interaction of rapid climate change and habitat fragmentation

conventions focused on the protection of the environment. Many of these have resulted in legislation as well as the creation of governing bodies to champion the cause of environmentalism.

The Conference on Human Environment held in Stockholm in June 1972 hosted 113 countries and was the first international forum to address environmental challenges on a global level. The resulting consensus of the conference was that there was a need for a common outlook and for common principles to inspire and guide the peoples of the world in the preservation and enhancement of the human environment. It was through this conference that the United Nations Environment Programme (UNEP) was founded.

In 1975 the Convention on International Trade in Endangered Species (CITES) of flora and fauna was signed on March 3, 1973 in Washington, DC. This Convention was to ensure that the survival of animals and plants was not threatened by international trade of endangered species. Member states promised to ensure that local legislation is in accordance with the guidelines provided by CITES.

In 1976, HABITAT was the first global meeting to link human settlements and the natural environment. It was during this meeting that world leaders started to recognize and debate the consequences of rapid urbanization, especially in the developing world. Members of the conference convened to deal with the perceived threat to the environment by human activity. The United Nations Human Settlements Programme, UN-HABITAT, was developed as a result of this conference.

In 1981, The World Health Assembly adopted a global strategy for health for all by the year 2000. It was agreed upon that member states would individually develop national plans, policies and programmes according to the guiding principles issued by the Executive Board of the World Health Organization (WHO) to ensure priority for public health at all levels of government.

The International Conference on Environment and Economics (OECD) in 1984 brought together world leaders as well as scientists and economic experts to discuss the environmental challenges and consequences of improving global economic conditions. Organisation for Economic Co-operation and Development (OECD) member countries agreed to develop appropriate environmental structures and policies to ensure environmental sustainability in the face of increasing economic development and cooperate to adopt more effective and efficient environmental policies.

The 1987 Montreal Protocol on Substances that Deplete the Ozone Layer was created to ensure member states of UNEP measure, control and eliminate substances harmful to the ozone layer such as chlorofluorocarbons (CFCs). Another conclusion of this Protocol was that new technologies relating to the elimination of substances harmful to the ozone layer would be shared among member states (UNEP, 2000).

In 1988 the IPCC, a scientific intergovernmental body, was tasked with evaluating the risk of anthropogenic climate change. Since then there have been four assessment reports released by the IPCC that summarize the latest knowledge surrounding the science of climate change, mitigation and adaptation strategies, and the political implications of future climate change.

The 1992 Earth Summit, the United Nations Conference on Environment and Development (UNCED), was held in Rio de Janeiro in Brazil, where 172 governments participated. Several issues were addressed during the summit that included: patterns of production (toxic waste), alternative sources of energy, reliance on public transportation systems in order to reduce vehicle emissions, and the growing scarcity of fresh water. During this conference the United Nations Framework Convention on Climate Change (UNFCCC) treaty was signed, which aims to stabilize greenhouse gas concentrations in the atmosphere at a level that would prevent dangerous anthropogenic interference with the climate system (UNFCC, 1992)

In the 1997 Kyoto Protocol, 159 nations attending the Third Conference of the Parties (COP-3) to the UNFCC in Kyoto, Japan agreed to reduce worldwide emissions of greenhouse gases. The Protocol also allows for several flexible mechanisms, such as emissions trading to help developed countries meet their greenhouse gas emission targets.

The 2009 Copenhagen Summit drafted the Copenhagen Accord, which recognized that climate change is one of the greatest challenges of the present day and that actions should be taken to keep any temperature increases to below 2°C.

Due to the last three decades of international governmental efforts and the work of numerous academic researchers, the concept of sustainability gained prominence. The next section briefly examines the key features of sustainability before exploring in more detail the notion of sustainable innovation.

SUSTAINABILITY

The term 'sustainability' has several definitions. The classic definition of sustainability, by the World Business Council for Sustainable Development (WBCSD), is 'meeting the needs of the current generation without jeopardizing future generations'. Another broader definition, stemming from John Ehrenfeld's (2009) book *Sustainability by Design*, states that, 'sustainability is the possibility that humans and other life will flourish on Earth forever'. Ehrenfeld identifies problematic cultural attributes (the unending consumption that characterizes modern life) and outlines practical steps toward developing sustainability as a mindset.

Our own definition of sustainability is: bringing about a balance between nature and humans for stable evolution. It implies balance between multigenerational needs over long-term goals, and balance between rich and poor nations. Furthermore, it is equally important that a balanced approach between production and consumption is established, where competing stakeholders and organizations function under this awareness.

SUSTAINABILITY CHALLENGES

There are four classic challenges to sustainability:

1. Population – by the year 2042, world population will reach 9.2 billion
    (Ehrlich and Ehrlich, 1991), which makes population a central chal-
    lenge of sustainability. Much of this growth will occur in India, China
    and other developing countries. With increasing population comes
    the need for increased production of goods and energy. With current
    production practices and technologies this increase will undoubt-
    edly come at the cost of environmental degradation. Thus arises the
    challenge of meeting the needs of an increasing population while
    maintaining a worldwide ecological–economic equilibrium. Measures
    are needed for stabilizing human populations and for reducing
    consumption.
2. Energy use – current patterns of producing and consuming energy
    are clearly unsustainable. Globally, we are overly dependent (80 per
    cent) on fossil fuel-based energy. Fossil (oil and natural gas) fuels are
    non-renewable, finite, and will be exhausted at some point. They also
    emit greenhouse gases that are the root cause of global warming and
    climate change. Therefore, the future energy pathway that we choose
    will need to move towards more sustainable and renewable sources of
    energy such as solar, wind and geothermal. In addition, we need to
    consider reducing the worldwide rate of energy consumption. Current
    consumption patterns are highly wasteful. Over half the energy pro-
    duced is lost in transmission losses, low-energy-efficiency equipment
    and wasteful habits. Global economic shifts towards renewable energy
    and energy conservation are necessary and a fundamental precursor
    for ecological sustainability.
3. Pollution – pollution in the form of greenhouse gases and wastes of
    all types is an increasing problem around the world. Minimizing pol-
    lution and encouraging efficient waste management must be at the
    forefront of sustainability initiatives. To achieve sustainability, indus-
    trial production must minimize the negative impact that production
    systems have on the environment by using cleaner production tech-
    nologies. This entails reducing pollution and minimizing toxic and
    solid wastes. These changes require increasing production efficiency,
    reducing technological hazards, and recycling and reusing materials.
4. Ecosystem resources – many global ecosystems (forests, fisheries,
    coral reefs, deserts, and their wildlife) are under the risk of extinction.
    The number of species that are now going extinct annually exceeds
    the number of new species formation as per natural rates of evolution

established over millions of years. As much as 15–37 per cent of species will be committed to extinction in 2050 from future climate changes (Thomas et al., 2004). Therefore, managing ecosystem resources and biodiversity to maintain their long-term viability is a priority. Global ecosystems are currently under tremendous stress from anthropogenic activities and climate change. This also implies conserving non-renewable natural resources (freshwater resources) and maintaining the integrity of sensitive ecosystems such as rain forests, deserts and marine habitats.

SUSTAINABLE INNOVATION

How do these challenges translate as issues for organizations? In this section I explore this issue with a focus on business organizations – corporations. For multinational corporations (MNCs) the way to think about sustainability and sustainable innovation is in terms of their approach to production and consumption. MNCs and other organizations are systems of vision, inputs, throughput and outputs (Figure 1.2).

Vision is comprised of the values, mission and strategy of the organization. Often, this initial phase of an organization will help to determine the required pathway to sustainable innovation. Inputs include the energy, raw materials and people of the organization that work to achieve

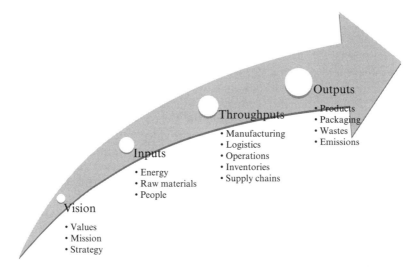

Figure 1.2   Multinational corporations are systems of vision, inputs, throughputs and outputs

the organizational vision. Throughputs include manufacturing, logistics, operations, inventories and supply chains. Outputs are the products, packaging, wastes and emissions created by the organization.

Innovation can occur in each of these elements. Vision and strategy can be imaginative and include multiple stakeholder and intergenerational long-term thinking. Inputs can be made sustainable through energy conservation, use of renewable or recycled materials, and sustainable human resource policies. Throughput systems can be made sustainable by using resource- and energy-efficient production processes, greening the supply chain, and efficient transportation and logistics systems. Outputs such as products and wastes can be made sustainable through eco-design of products and services, and waste management practices. Some examples of simple sustainable innovations are provided below.

In 2005, General Electric (GE) launched a programme called 'Ecomagination', intended to position itself as an environmentally aware company. GE is currently one of the biggest players in the wind power industry, and it is also developing new environment-friendly products such as hybrid locomotives, desalination and water reuse solutions, and photovoltaic cells. The company has set goals for its subsidiaries to lower their greenhouse gas emissions and develop sustainable solutions for the future. They have also invested in fuel cells, lower-emission aircraft engines, lighter and stronger durable materials, and efficient lighting. Since entering the renewable energy industry in 2002, GE has invested more than $850 million in renewable energy technology.[1]

The 3M Corporation adopted a 'Pollution Prevention Pays Program', also known as 3P, in 1975, which aims to prevent pollution at source, in products and manufacturing. 3P has resulted in the elimination of more than 3 billion pounds weight of pollution and saved nearly $1.4 billion. 3M encouraged its employees to come up with ideas for saving costs and reducing pollution; in turn, the 3M company would fund these ideas. Simple commonsense solutions provided innovative changes to processes, systems and practices. For example, collecting excess spray paint from a production line with a specially designed suction funnel limits cleaning requirements and reduces health-hazardous working conditions. In addition, the funnel collects the excess paint and recycles it. With an underlying concept based on the belief that a preventative approach is more effective, technically sound and economical than conventional pollution controls, 3M has become a leader in sustainable innovation.[2]

United Parcel Service (UPS) has a large truck fleet of 95000 vehicles for delivery of mail and parcels. Even a small saving in fuel gets multiplied a thousand times over when implemented at the fleet level. When challenged to reduce CO_2 emissions and fuel costs, engineers came up with a surprisingly simple innovation – eliminate left turns. Apparently in countries which drive on the right, left turns, on average, take 10 to 40 seconds more of engine idling than right turns. Using geographical information systems mapping and optimization software, transportation analysts were able to design delivery routes in the USA while minimizing left turns. In 2007, this resulted in a reduction of 30 million miles of driving, 3 million gallons of gasoline, and 32 000 tons of emissions.[3]

The trucking industry faced a challenge of idling trucks at rest stops. Truck drivers are required by law to rest after a designated period. Most of them pull over to a rest stop and leave their engines idling for the rest period to heat or cool their truck cabin. In other words, a 500-horse power engine is running to power a small heating/cooling unit. An industry study identified that a simple auxiliary generator with a 20 horsepower engine could do the same job. The innovative engine substitution saves approximately $6500 per year per truck.[4]

The Montreal Protocol for elimination of ozone-depleting substances was signed in 1987. Companies using CFCs were mandated to eliminate their use over a ten-year period. One of the largest users of CFCs was the electronics industry. It used CFCs to clean printed circuit boards. A third of all consumption of CFCs was in this application. The industry set up a task force led by Nortel Networks to find a solution to this challenge. Within six months they discovered that printed circuit boards could be cleaned equally effectively by replacing CFCs by a mixture of an industrial detergent and lemon rind, with some minor modifications in the cleaning procedures. This innovation led to an industry-wide elimination of CFCs within a year (Berry and Rondinelli, 1998).

Throughout daily office work, typing up reports and creating word documents is a common activity and printing these documents is often a necessary outcome. However, different font styles require different amounts of ink to print. In other words, simply changing the font style, to one that uses less ink when printing, results in ecological and economical savings. A recent study by Print.com showed that normal fonts use up to 60 per cent more ink per page than an eco-efficient font. Century Gothic is one of the most ink-efficient fonts in the standard set of fonts in word processors. There are also specially designed eco-fonts.[5] By selecting ecologically efficient fonts offices can reduce the use of ink, reduce the number of cartridges that go to landfill, and save costs. Average savings can be as high as $80 per year per printer.

IBM uses eco-innovation to create a business service, 'Big Green'. This service creates energy-efficient server farms and building controls.

A server farm is a collection of computer servers usually maintained by an enterprise to accomplish server needs far beyond the capability of one machine. According to the US Department of Energy, data centre energy consumption doubled from 2000 to 2006, reaching more than 60 billion kilowatt hours per year. A team of engineers has successfully tested a novel system that greatly improves the efficiency of data centre cooling. Through a combination of low-energy-consuming hardware and smart software building, energy costs can be reduced by 10 to 20 per cent in most buildings. Given that buildings consume 60 per cent of all energy, the potential for energy savings is enormous.[6]

In industrial ecosystems, organizational outputs can be made sustainable through reuse and recycling. Products, services, wastes, packaging and buildings can be deliberately reused to save materials, fuel and energy, and reduce emissions. For example, the notion of industrial ecosystems experiments with waste management practices that recycle, reuse, recover and regenerate, to make one company's waste the raw material for another. One famous example is a power plant in Kalundborg, Denmark. The coal ash from the power plant is sent for use as raw material in a Wall Board plant. Sulphurous oxides are trapped from gas emissions and sent to a sulphuric acid plant as inputs. Waste steam from the power plant is sent to local fisheries to heat ponds and to the city of Kalundborg for heating buildings. Waste from both the fisheries and the city is burnt as fuel in the power plant. The collective eco-footprint and water use has been brought down significantly within this industrial ecosystem.[7] Figure 1.3 illustrates the industrial ecosystem at Kalundborg.

The famous 'eat fresh' sandwich restaurant chain, Subway, has taken proactive steps to attaining sustainability throughout its restaurants and has become more environmentally responsible for its distribution, packaging and construction activities. One of the first steps was to redesign the company's supply-chain operation, which saves the company about 1.6 million gallons of diesel fuel a year. It also switched to 100 per cent recyclable napkins, helping to save an estimated 140 000 trees annually. By redesigning its shipping packaging, it cut out more than 97 000 pounds of plastic a year. More importantly, the franchise worked closely with the US Green Building Council to design and build a store that conserves energy and water and meets the highest green building and performance standards. Its 'Eco-Stores' have low volatile organic compound (VOC) paints and sealants, wood trim from managed forests, energy-efficient LED track lighting and pendant lights with compact fluorescent bulbs. Their windows are film-coated to block ultraviolet light, and solar tubes on the roof convert sunlight for auxiliary lighting. During their construction process Subway used LEED (Leadership in Energy and

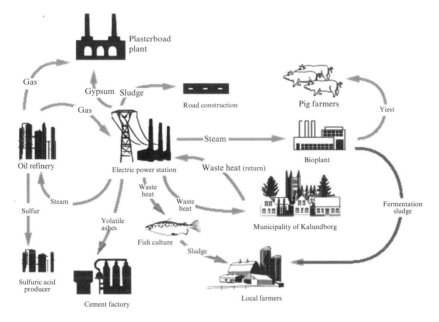

Plasterboad plant

Gas

Gypsum Sludge

Gas

Road construction

Pig farmers

Yiest

Oil refinery

Electric power station

Steam

Waste heat (return)

Bioplant

Sulfur

Steam

Waste heat

Waste heat

Municipality of Kalundborg

Fermentation sludge

Volatile ashes

Fish culture

Sludge

Sulfuric acid producer

Cement factory

Local farmers

Source: *Ecodecision* (Spring 1996: 20)

Figure 1.3 *Industrial ecosystem at Kalundborg, Denmark*

Environmental Design)-certified consultants to ensure sustainable building procedures.[8]

Hotels play a crucial role in environmental sustainability because they consume significant energy and water, and have large ecological footprints. As a result, Hilton Hotels has set out a framework of sustainability goals in order to reduce its environmental impact and become more energy efficient. Such goals include reducing its energy consumption, waste and CO_2 emissions by 20 per cent, as well as water consumption by 10 per cent. To measure these reductions Hilton has implemented the LightStay measurement system, which takes into account energy, water use, waste and carbon outputs associated with building operations and services provided at Hilton properties. The system measures indicators across 200 operational practices including housekeeping, paper product use, food waste, chemical storage, air quality and transportation. By the end of 2011, all 3500 properties within Hilton's global portfolio of brands were expected to use LightStay, making the company the first major multi-brand company in the hospitality industry to require property-level measurement of sustainability.[9]

From these examples sustainable innovation (SI) can be character-ized as changes in enterprise systems, procedures, and technology that bring about cost savings by reducing carbon footprints, improving social efficiencies and creating human-scale operations. Reducing one's carbon footprint entails reducing greenhouse gas emissions through energy effi-ciency, while social efficiencies refer to expansion of the value chain beyond economic benefits to social benefits for stakeholders. In other words, social efficiencies are measured not in economic terms but rather in terms of the social benefits that are created via a sustainable carbon innovation pathway. Lastly, human-scale operations are those that can be controlled transparently, accountably and responsibly by organizations and their stakeholders.

Similarly, eco-innovation (EI) refers to products and processes that contribute to sustainable development. Although SI is mostly used when referring to product or process design, and therefore focuses more on the technological aspects (as discussed above), both SI and EI focus on the commercial application of knowledge to elicit direct or indirect ecological improvements. In general, EI is all-encompassing and tends to consider the societal or political aspects of sustainability.

SI often emerges from bottom-up initiatives. Employees or users who have intimate knowledge of the product or service can conceive of improvements in existing methods. They may be more open to new ideas and not have vested interests in maintaining the status quo. Companies can uncover these ideas through 'deep listening' and subsidize new projects that are based on SI.

Perhaps the most misunderstood concept of sustainable innovation is that it does not involve radical new technologies but rather simple, low-technological improvements, which are made incrementally throughout the company. It relies on local investments and local knowledge from the people who understand the product best. More importantly, both economic and ecologic spheres are measured equally within the context of profit or loss. This is attained through mindful usage, where organizations are aware of the natural resource intensity of technologies over generations. At the same time, innovations made at the production level benefit from the savings of an entire production line. Every case is specific to the needs of the company and the clientele, demanding specific pressure depending on the situation. In general, SI relies on systemic views and not on the fragmentation of parts. In other words, the innovation is taken into consideration under a total systems approach, where its improvements are measured throughout the entire organization, production and institutional space.

UNLEASHING INNOVATION

In this chapter I have offered some ways of thinking about sustainable innovation as a response to global climate change. This preliminary analysis does not offer any firm conclusions, but it does suggest some ways of thinking about and unleashing innovation in organizations. A schematic for thinking about sustainable innovation is presented in Figure 1.4.

There are several precursors to unleashing and maintaining sustainable innovation within organizations. First and foremost is creating an environment or climate for innovation. Innovation requires a certain level of freedom, playfulness, and empowerment of employees and other stakeholders. It does not prosper in highly regimented and conformance-seeking environments. Evoking playfulness among a company's employees entails engaging people at an emotional level with spirit and passion. It also involves giving employees the freedom to play with new ideas and to take risks, without penalty for failure.

It is important to empower people by giving them decision-making opportunities and the resources needed to implement their ideas. Creative ideas for sustainable innovation often occur where employees interact

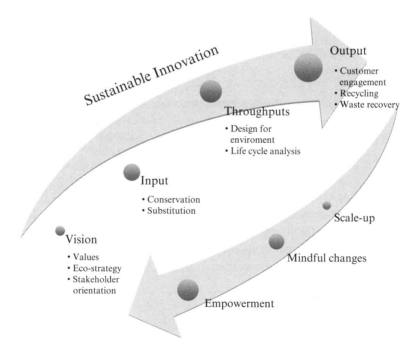

Figure 1.4   A schematic for thinking about sustainable innovation

with information systems, operating equipment and procedures. Often, these are lower-level employees in organizations, who do not have authority and power to implement their ideas. Organizations need to empower these employees to implement sustainability ideas (Melville, 2010).

Sustainable innovations emerge from mindful application of logical and sensible modification to operational systems to their present circumstances. Many organizational systems become ossified over time, by bureaucratic procedures, or traditional ways of thinking about and doing things that may no longer be relevant. Often, there are good historical reasons for these procedures to be in place. Mindful changes involve updating and renovating operational processes to adjust to changes in the environment and availability of new technologies

It is critical that organizations create a climate that encourages experimentation, where employees can mindfully apply novel solutions, make errors and learn from them. At the heart of sustainable innovation is the idea of creativity. Sustainable innovations are new ways of doing old things. They are based on creative insights. Creativity requires experimentation – trial and error. It cannot occur in conditions of rigid, formulaic and formatted environments. So organizations need to create a space for flexibility and experimentation.

Finally, companies must be willing to bend their 'business model' at times to scale up innovation across the organization. They must adapt to new innovations and incorporate behavioural and financial changes into their new business structures. The business model here refers to the basic logic of how a company generates its revenues and controls its costs – it is the core logic of the business and probably one of the most difficult and risky things to change. Some sustainable innovations require changing that core logic of the business. Failure to do so could result in missed opportunities for fundamental restructuring. For example, the Internet provided an opportunity for radical innovation in the newspaper industry. But the industry was set on its traditional business model (paper-based, reporter-generated content, editorial-controlled quality, paid subscriptions). It failed to incorporate the Internet innovations of digital delivery, blog reporting, self-editing and peer quality control in meaningful ways. As a result numerous newspapers have gone out of business.

CONCLUSION

The threat of climate change is upon us and a new restructuring of consumption and production habits is necessary to avoid environmental destruction. Sustainability theory and practices, which are deeply rooted and founded through international treaties and conventions, are a logical solution or response to climate change. By meeting the challenges of sustainability and implementing sustainable innovation practices throughout the production chain among multinational corporations it is possible to create a sustainable future. Simple small-scale technologies coupled with bottom-up initiatives from employee members can help companies unleash sustainable innovation. Ultimately, companies must be willing to bend their business model and accept change, as an opportunity to rethink current practices and design new environmentally sound, efficient, and sustainable measures.

NOTES

1. General Electric's 'Ecoimagination' programme, available at http://www.ecomagina tion.com (accessed 7 June 2012).
2. 3M Corporation's 'Pollution Prevention Pays Program', available at http://solutions.3m.com/wps/portal/3M/en_US/3MSustainability/Global/Environment/3P/ (accessed 4 April 2012).
3. Source: http://www.pressroom.ups.com/Fact+Sheets/UPS+Saves+Fuel+and+Reduc es+Emissions+the+%22Right%22+Way+by+Avoiding+Left+Turns (accessed 15 May 2012).
4. Source: http://www.epa.gov/smartway/documents/apu.pdf (accessed 6 February 2012).
5. More ecofonts are available at http://www.ecofont.com (accessed 6 June 2012).
6. IBM's 'Big Green' available at: http://www.ibm.com/ibm/green/index.ohtml (accessed 9 April 2012).
7. Source: http://www.bsdglobal.com/viewcasestudy.aspx?id=77 (accessed 6 June 2012).
8. Source: http://www.subway.com/subwayroot/AboutSubway/HelpingSociety/GoingGre en/index.aspx (accessed 3 March 2012).
9. Source: http://www.hiltonworldwide.com/aboutus/sustainability.htm (accessed 12 May 2012).

REFERENCES

Berry, M.A. and D.A. Rondinelli (1998), 'Proactive corporate environmental management: a new industrial revolution', *Academy of Management Executive*, **12**(2): 38–50.

Bruntland, G. (ed.) (1987), *Our Common Future: The World Commission on Environment and Development*, Oxford: Oxford University Press.

Cramer, W., A. Bondeau, F.L. Woodward et al. (2001), 'Global response of terrestrial ecosystem structure and function to CO_2 and climate change: results from six dynamic global vegetation models', *Global Change Biology*, **7**(4): 357–73.

Ehrenfeld, J.R. (2009), *Sustainability by Design: A Subversive Strategy for Transforming Our Consumer Culture New Haven*, New Haven, CT: Yale University Press.

Ehrlich, P. and A. Ehrlich (1991), *The Population Explosion*, New York: Touchstone.

Fox, C. et al. (2009), 'Transregional linkages in the north-eastern Atlantic – an "end-to-end" analysis of pelagic ecosystems', *Oceanography and Marine Biology: An Annual Review*, **47**: 1–75.

Goberville, E., G. Beaugrand, B. Sautour, P. Treguer and SOMLIT Team (2010), 'Climate-driven changes in coastal marine systems of Western Europe', *Marine Ecology – Progress Series*, **408**: 129–48.

Haanaes, K., M. Reeves, B. Balagopal, D. Arthur and M.S. Hopkins (2011), 'First look: the second annual sustainability and innovation survey', *MIT Sloan Management Review*, **52**(2): 77–83.

IPCC (2007), *Climate Change 2007: Synthesis Report. Contribution of Working Groups I, II and III to the Fourth Assessment Report of the Intergovernmental Panel on Climate Change*, R.K. Pachauri and A. Reisinger (eds), Geneva.

Jump, A.S. and J. Penuelas (2005), 'Running to stand still: adaptation and the response of plants to rapid climate change', *Ecology Letters*, **8**(9): 1010–20.

Morin, X., M.J. Lechowicz, C. Augspurger, J. O'Keefe, D. Viner and I. Chuine (2009), 'Leaf phenology in 22 North American tree species during the 21st century', *Global Change Biology*, **15**(4): 961–75.

Nelson, R.R. and S.G. Winter (1982), *An Evolutionary Theory of Economic Change*, Cambridge, MA, USA and London, UK: Belknap Press.

Science Academies (2005), 'Joint statement of Science Academies of G8 plus Brazil, Russia, India and China – global response to climate change, 2005'.

Shrivastava, P. (1995), 'The role of corporations in achieving ecological sustainability', *Academy of Management Review*, **20**(4): 936–60.

Stern, N. (2007), *The Economics of Climate Change: The Stern Review*, Cambridge: Cambridge University Press.

Thomas, C. et al. (2004), 'Extinction risk from climate change', *Nature*, **427**: 145–8.

UNFCCC (1992), *United Nations Framework Convention on Climate Change*, Chatelaine, Switzerland: UNEP/IUC for the Climate Change Secretariat.

Weller, G. (1998), 'Regional impacts of climate change in the Arctic and Antarctic', *Annals of Glaciology*, **27**(27): 543–52.

Web Reference

Nidumolu, R., C.K. Prahalad and M.R. Rangaswami (2009), 'Why sustainability is now the key driver of innovation', *Harvard Business Review*, available at http://hbr.org/2009/09/why-sustainability-is-now-the-key-driver-of-innovation/es (accessed 6 June 2012).

2. Understanding eco-innovation for enabling a green industry transformation

Tomoo Machiba

INTRODUCTION

The Organisation for Economic Co-operation and Development (OECD) Green Growth Strategy was launched in May 2011 (OECD, 2011a)[1] to substantiate the Declaration on Green Growth adopted by at the OECD Council Meeting at Ministerial Level (MCM) two years earlier (OECD, 2009a). Together with the Green Economy Report compiled by the United Nations Environment Programme (UNEP, 2011), the launch of this Strategy is hoped to bring out sustainable development under the spotlight of international politics once again as nearly 20 years have passed since the international community adopted this new concept of progress at the Earth Summit in Rio de Janeiro.

This OECD Strategy generally defines green growth as 'fostering economic growth and development while ensuring that natural assets continue to provide the resources and environmental services on which our well-being relies' (OECD, 2011a: 18). It implies that policies should encourage green investment in order to simultaneously contribute to economic recovery in the short term and help to build the environmentally friendly infrastructure required for a green economy in the long term. Industries around the world, particularly in OECD countries, are already showing a greater interest in sustainable production and have been undertaking a number of corporate social responsibility (CSR) initiatives during the last decade. However, progress falls far short of meeting the pressing global challenges such as climate change, energy security and depletion of natural resources.

The political and economic challenges for OECD countries are daunting and incremental improvement within the current technology regime is not enough to meet such challenges. Industry needs to be restructured and existing and breakthrough technologies must be more innovatively applied to realise green growth. The OECD Directorate for Science, Technology and Industry (DSTI) is thus aiming to contribute to the implementation of the OECD Strategy by promoting the role of innovation for realising green growth. Raising efficiency in resource and energy use and engaging in a broad range of innovations to improve environmental performance will help countries to create new industries and jobs in coming years. The ongoing economic crisis and negotiations to tackle climate change should be seen as an opportunity to shift to a greener economy.

This chapter presents part of the interim outcomes from the OECD project on Green Growth and Eco-innovation.[2] It mainly provides a theoretical underpinning of eco-innovation, takes stock of the existing knowledge in this area and proposes a new perspective for more comprehensive analyses in the future. Firstly, this chapter outlines a policy context of green growth and eco-innovation and attempts to develop a conceptual framework of eco-innovation based on the existing innovation research for a common understanding. Secondly, the framework is applied to understand the evolution of corporate activities for sustainable production and some good practices from the existing literature. Thirdly, the chapter envisions the potential of diverse approaches of eco-innovation captured by the framework, with a particular focus on the role of business models that enable a successful uptake of innovative products and services. Lastly, the role of diverse policy instruments in the acceleration of eco-innovation is reviewed and the chapter concludes by setting the agenda for the future eco-innovation research.

GREEN GROWTH AS NEW POLICY CROSSROADS

'Green growth' and a 'green economy' by no means replace sustainable development, but can be considered as vital endeavours to set a clearer and more focused policy agenda for delivering on a number of the key aspirations that the Rio Summit set as the Agenda 21. Despite its wide acceptance and the development of national strategies and many related initiatives, sustainable development and the fellow concept of 'sustainable consumption and production', which was elevated as a focused agenda following the 2002 World Summit on Sustainable Development (WSSD) in Johannesburg, have not been effective to tackle climate change nor make a paradigm shift in the course of economic and social development. This may be partly because a wide range of aspects in three dimensions – economic, social and environmental – have been embraced under the sustainability concept, but the synergies between three pillars have not been fully pursued, keeping the policy silos very much intact. The time is ripe to consider a new strategy to revive the sustainability agenda and to integrate it practically into mainstream policies as well as industry activities.

Meanwhile, the continuing financial and economic crisis since late 2008 created room for public policies aimed at encouraging economic recovery to tap new sources of growth on more environmentally and socially sustainable grounds. Also stimulated by rising oil prices, energy insecurity and challenges of climate change, the consideration of 'green stimulus' measures has been providing a unique opportunity for policy-makers from different fields to deliberate how to address economic and environmental sustainability simultaneously.

Green growth ultimately implies a condition that growth does not lead to overall depletion of natural capital, or vice versa. In other words, green growth generally entails decoupling economic growth from environmental degradation. More precisely, it implies an economy that enables us to reduce resource use per unit of value added (relative decoupling) or, furthermore, an economy that keeps resource use and environmental impacts stable or declining while the overall growth rate is positive (absolute decoupling). For OECD countries, the challenge today is increasingly about achieving an absolute decoupling, while in many developing economies the challenge is to achieve gains in living standards without imposing excessive burdens on environmental carrying capacity.

Over the last few decades, OECD countries have been able to achieve an absolute decoupling between growth of gross domestic product (GDP) and emissions of certain acidifying substances such as sulphur oxides (SOx) and nitrogen oxides (NOx), while they were also able to achieve a relative decoupling between GDP growth and greenhouse gas (GHG) emissions. In reality, however, the state of OECD countries is far from green growth, let alone the state of non-OECD economies. In many areas, environmental pressures have continued to rise as OECD economies have grown and GHG emissions also have continued to rise (OECD, 2011b). Without new policy action, an International Energy Agency (IEA) analysis suggests that global energy-related GHG emissions are likely to double by 2050, whilst the G-8 leaders agreed to aim for halving global emissions during the same period (IEA, 2010). Importantly, improvements in efficiency have often been offset by increasing consumption and outsourcing, while efficiency gains in some areas are outpaced by scale effects.

The challenges apparently cannot be met by 'business as usual', as with existing production technology and consumer behaviour, positive outcomes can only be produced up to a point. Significant innovation – in both the creation of new technologies, products and processes, as well as their application and diffusion – will be required to start reducing the global GHG emissions and deliver the decoupling. Innovation also helps deliver the objectives at the least possible cost and thus should play a key role in greening economy.

INNOVATION FOR GREEN GROWTH

As Michael Porter claimed that 'innovation is the central issue in economic prosperity', it is now well recognised that innovation is a driver of economic and social progress on the national (macro) level as well as a driver of business success and competitive advantage at the firm (micro) level (Porter and van der Linde, 1995). Recently, increasing attention has been paid to innovation as a way for industry and policy-makers to achieve more radical improvements in corporate environmental practices and performance. If countries want to move towards a more ecologically sound and prosperous society, it is important to promote the right kinds of innovation. Such innovation should allow for new ways of addressing current and future environmental problems and decreasing energy and resource consumption, while promoting sustained economic activities. This type of innovation is referred to as eco-innovation (or green innovation). Many companies started to use this or similar terms to describe their contributions to sustainable development. A few governments are also promoting the concept as a way to meet sustainable development targets while keeping industry and the economy competitive. However, while the promotion of eco-innovation by industry and government generally involves the pursuit of both economic and environmental sustainability, the scope and application of the concept tend to differ.

In the European Union (EU), eco-innovation is considered to be an important contributor to the wider objectives of its Lisbon Strategy for competitiveness and economic growth, and a key factor of green growth in the recent Europe 2020 strategy. The concept is promoted primarily through the Eco-Innovation Action Plan (formerly the Environmental Technology Action Plan), which defines eco-innovation as 'the production, assimilation or exploitation of a novelty in products, production processes, services or in management and business methods, which aims, throughout its lifecycle, to prevent or substantially reduce environmental risk, pollution and other negative impacts of resource use (including energy)'.[3] Environmental technologies are also considered to have promise for improving environmental conditions without impeding economic growth in the United States, where they are promoted through various public–private partnership programmes and tax credits (OECD, 2008).

To date, the promotion of eco-innovation has focused mainly on environmental technologies, but there is a tendency to broaden the scope of

Target Field	Industry		Social infrastructure		Personal lifestyle
	Manufacturing	Service	Energy	Transportation/urban	
Technology	Sustainable manufacturing Innovative R&D (energy saving etc.) Green ITC Rare metal recycling	Innovative R&D Building Energy (Management System)	Innovative R&D (renewable energy, batteries) Superconducting transmission	Innovative R&D (intelligent transport systems) Green automobiles Maglev	Heat pump
Business model	Green procurement (including BtoB) Green servicizing EMA LCA	Energy services Environmental rating/green finance	Green certification	Modal shift	Green procurement Cool biz Green finance
Societal system (institution)	Environmental labeling system Starmark Green investment		Top Runner Programme PRS Act (Renewables Portfolio Standard)	Green tax for automobiles Next generation vehicle and fuel initiative (METI)	Telework telecommuting Work-life balance

Source: Japanese Ministry of Economy, Trade and Industry (METI)

Figure 2.1 The scope of Japan's eco-innovation concept

the concept. In Japan, the government's Industrial Science Technology Policy Committee defined eco-innovation as 'a new field of techno-social innovations [that] focuses less on products' functions and more on [the] environment and people' (METI, 2007: 56). Eco-innovation is there seen as an overarching concept which provides direction and vision for pursuing the overall societal changes needed to achieve sustainable development (Figure 2.1).

The OECD is primarily studying innovation following the OECD/Eurostat Oslo Manual for the collection and interpretation of innovation data. This manual describes innovation as 'the implementation of a new or significantly improved product (good or service), or process, a new marketing method, or a new organisational method in business practices, workplace organisation or external relations' (OECD and Eurostat, 2005: 46). This provides a good overview on where innovation occurs beyond technology spheres but does not shed enough light on how it occurs and what it is developed for, which are essential to understand the nature of eco-innovation as it particularly concerns the scope of changes and the impact the changes can create for improving environmental conditions.

Charter and Clark (2007: 10) provide an alternative useful classification of eco-innovation based on the levels of making differences from the existing state, as shown below:

- Level 1 (incremental): incremental or small, progressive improvements to existing products.
- Level 2 (redesign or 'green limits'): major redesign of existing products (but limited level of improvement that is technically feasible).
- Level 3 (functional or 'product alternatives'): new product or service concepts to satisfy the same functional need, for example teleconferencing as an alternative to travel.
- Level 4 (systems): design for a sustainable society.

In addition to the above two different dimensions of innovation framed by the OECD and Charter and Clark, the concept of eco-innovation entails two other significant, distinguishing characteristics from that of ordinary innovation. Firstly, eco-innovation includes both environmentally motivated innovations and unintended environmental innovations. The environmental benefits of an innovation may be a side-effect of other goals such as reducing costs for production or waste management (Kemp and Pearson, 2008). In short, eco-innovation is essentially innovation that reflects the concept's explicit emphasis on a reduction of environmental impact, whether such an effect is intended or not. Secondly, eco-innovation should not be limited to innovation in products, processes, marketing methods and organisational methods, but also includes innovation in social and institutional structures (Rennings, 2000). Eco-innovation and its environmental benefits go beyond the conventional organisational boundaries of the innovator and affect the broader societal context through changes in social norms, cultural values and institutional structures.

Built upon the above OECD/Eurostat definition, eco-innovation can therefore be defined as 'the implementation of new, or significantly improved, products (goods and services), processes, marketing methods, organisational structures and institutional arrangements which, with or without intent, lead to environmental improvements compared to relevant alternatives' (OECD, 2010a: 40). The EU's Eco-Innovation Observatory (EIO) project similarly defines eco-innovation as 'the introduction of any new or significantly improved product (good or service), process, organisational change or marketing solution that reduces the use of resources and decreases the release of harmful substances across the whole life-cycle'.[4]

Synthesising the above existing understanding and definitions,

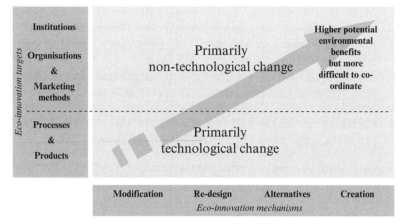

Figure 2.2   A proposed framework of eco-innovation

eco-innovation can be analysed from three dimensions, namely in terms of an innovation's target, mechanism and impact. Figure 2.2 presents an overview of eco-innovation and its typology:

'Target' refers to the basic focus of eco-innovation. Following the OECD/Eurostat Oslo Manual, the target of an eco-innovation may be:

- Products, involving both goods and services.
- Processes, such as a production method or procedure.
- Marketing methods, for the promotion and pricing of products, and other market-oriented strategies.
- Organisations, such as the structure of management and the distribution of responsibilities.
- Institutions, which include the broader societal area beyond a single organisation's control, such as institutional arrangements, social norms and cultural values.

The target of the eco-innovation can be technological or non-technological in nature. Eco-innovation in products and processes tends to rely heavily on technological development; eco-innovation in marketing, organisations and institutions relies more on non-technological changes (OECD, 2007).

'Mechanism' relates to the method by which the change in the eco-innovation target takes place or is introduced. It is also associated with the underlying nature of the eco-innovation – whether the change is of a technological or non-technological character. Four basic mechanisms are identified:

- Modification, such as small, progressive product and process adjustments.
- Redesign, referring to significant changes in existing products, processes, organisational structures, and so on.
- Alternatives, such as the introduction of goods and services that can fulfil the same functional need and operate as substitutes for other products.
- Creation: the design and introduction of entirely new products, processes, procedures, organisations and institutions.

'Impact' refers to the eco-innovation's effect on the environment, across its lifecycle or some other focus area. Potential environmental impacts stem from the eco-innovation's target and mechanism and their interplay with its socio-technical surroundings. Given a specific target, the potential magnitude of the environmental benefit tends to depend on the eco-innovation's mechanism, as more systemic changes, such as alternatives and creation, generally embody higher potential benefits than modification and redesign.

UNDERSTANDING INDUSTRY PRACTICES THROUGH THE ECO-INNOVATION FRAMEWORK

Industries have traditionally addressed pollution concerns at the point of discharge. Since this end-of-pipe approach is often costly and ineffective, industry has increasingly adopted 'cleaner production' by reducing the amount of energy and materials used in the production process. Many firms are now considering the environmental impact throughout the product's lifecycle and are integrating environmental strategies and practices into their own management systems. Some pioneers have been working to establish a 'closed-loop production' system that eliminates final disposal by recovering wastes and turning them into new resources for production, as exemplified in remanufacturing practices and eco-industrial parks.

This evolution of sustainable production initiatives can be viewed as facilitated by eco-innovation and classified according to the dimensions proposed in the previous section. Figure 2.3 provides a simple illustration of the general conceptual relations between sustainable production and eco-innovation. The steps in sustainable production are depicted in terms of their primary association with respect to eco-innovation facets. While more integrated initiatives such as closed-loop production can potentially yield higher environmental improvements in the medium to long term, they can only be realised through a combination of a wider range of inno-

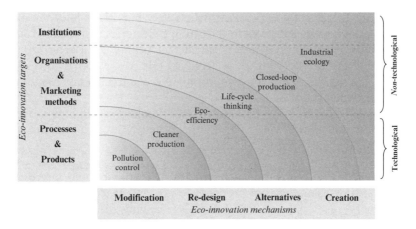

Figure 2.3 Conceptual relationships between sustainable production and eco-innovation

vation targets and mechanisms and therefore cover a larger area of this figure.

For instance, an eco-industrial park cannot be successfully established simply by locating manufacturing plants in the same space in the absence of technologies or procedures for exchanging resources. Process modification, product design, alternative business models and the creation of new procedures and organisational arrangements need to go hand in hand to leverage the economic and environmental benefits of such initiatives. This implies that as sustainable production initiatives advance, the nature of the eco-innovation process becomes increasingly complex and more difficult to coordinate.

Although such advanced initiatives may have their sources in technological advances, technology alone cannot make a great difference. It has to be associated with organisational and social structures and with human nature and cultural values. While this may indicate the difficulty of achieving large-scale environmental improvements, it also hints at the need for industries to adopt an approach that aims to integrate the various elements of the eco-innovation process so as to leverage the maximum environmental benefits. The feasibility of their eco-innovative approach would depend on the organisation's ability to engage in such complex processes.

While incremental optimisation through energy saving and eco-efficiency measures is playing an important role in widely diffusing greener practices, 'it's the disruptive end of the eco-innovation spectrum that is the most promising in the long term' (Future Think, 2008; Hellström, 2007). More radical forms of eco-innovation become key to enable a sustainable transition. Incremental improvements are very important but they may also help lock social practices into existing trajectories and make radical solutions, which require changes in the current technological or infrastructural regime, more difficult to be deployed (Hellström, 2007). Investing in radical solutions is therefore important and could maximise long-term gains and wider impacts.

To understand the basic mechanisms of different greening options from the innovation point of view, a distinction between three types of innovations can be made (Scrase et al., 2009):

- Incremental innovation, which aims at modifying and improving existing technologies or processes to raise efficiency of resource and energy use, without fundamentally changing the underlying core technologies. Surveys of innovation in firms demonstrate that this is the dominant form of innovation and eco-innovation in industry.
- Disruptive innovation, which changes how things are done or specific functions are fulfilled, without necessarily changing the underlying technological regime itself. Examples include the move from manual typewriters to word processors, or the change from incandescent to fluorescent lighting.
- Radical innovation, which involves a shift in the technological regime of an economy and can lead to changes in the economy's enabling technologies. This type of innovation is often complex and is more likely to involve non-technological changes and mobilise diverse actors. Radical innovations could include not only the development of radical, breakthrough technologies but also a reconfiguration of product-service systems, for example, by closing the loop from resource input to waste output ('cradle to cradle'), and the building of business models that reshape the way consumers receive value on the one hand and reduce material use on the other.

A sophisticated combination of these different types of innovation, together with new organisational and managerial arrangements, could bring out far-reaching changes in the techno-social system and enable a long-term green transformation by affecting several branches of the economy, including consumers. One such example is the introduction of a new urban mass-transit system which could be realised through a combination of changes to control systems (as facilitated by communications technologies), organisational practices (such as moves from hierarchical to networked collaboration), infrastructure management (such as those enabled by computing technologies), environmental monitoring (pushed by advances in remote sensing), manipulation techniques (as in genomics) or materials production (such as those made possible by modern industrial chemistry and nanotechnology) (Steward, 2008; Scrase et al., 2009).

Such systemic (or transformative) innovation is more likely to take place far beyond the boundaries of one company or organisation, as it often requires the transformation, replacement or establishment of complementary infrastructures. It is the innovation characterised by fundamental shifts in how society functions and how its needs are met (Geels, 2005). From the perspective of the transition to a greener economy and the decoupling of growth and environmental impacts, there is a growing attention towards systemic innovation as it could bring wider and persistent impacts in the medium to long term (OECD, 2011b; Smith, 2008; Scrase et al., 2009). Systemic eco-innovation is increasingly considered a cornerstone for the green economy as it may help the society to exit the current hydrocarbon-based technology regimes.

However, systemic innovation also 'involves substantive risky investments by its champions, conflicts between emergent and incumbent actors, and reconfiguring traditional sectoral and policy boundaries' (Steward, 2008: 15). One of the imperative conditions for such innovation is social and cultural changes, adopting new values and behaviour on both the producer and the consumer sides. The changes which systemic innovation brings would be difficult to predict and direct and would not be in linear processes, while the society needs to steer innovation into a direction of decreased environmental burden.

Figure 2.4 provides a conceptual distinction between incremental, disruptive and radical and systemic eco-innovation. Although drawing boundaries between different levels of eco-innovation activity is not

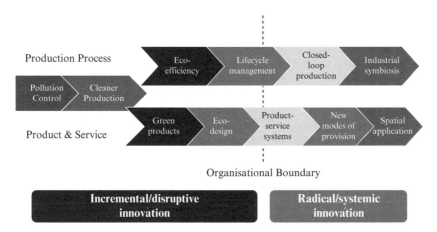

Figure 2.4   Diverse types of eco-innovation: incremental to systemic

necessarily easy, it can generally be considered that radical and systemic eco-innovations include those on the right-hand side of the figure. It should also be noted that incremental and disruptive innovation is in fact sometimes part of, or even a prerequisite for more radical and systemic changes.

APPLYING THE ECO-INNOVATION FRAMEWORK FOR GOOD PRACTICES

To better understand current applications of eco-innovation, a small sample of sector-specific examples from manufacturing industries were reviewed in light of the above framework (Figure 2.2). Examples from three sectors were chosen for this preliminary review: (1) the automotive and transport industry; (2) the iron and steel industry; and (3) the electronics industry. The examples draw mainly on the interaction with industry practitioners made through the OECD project (Table 2.1). The examples are not meant to represent best practices but were selected to illustrate the diversity of eco-innovation, its processes and the different contexts of its realisation.[5] Following is an overview of the examination of each sector's general practices and examples based on the proposed eco-innovation framework.

Table 2.1   Industry practices examined through the eco-innovation framework

Industry and company/association	Eco-innovation example
Automotive and transport industry	
The BMW Group	Improving energy efficiency of automobiles
Toyota	Sustainable plants
Michelin	Energy saving tyres
Vélib'	Self-service bike sharing system
Iron and steel industry	
Siemens VAI, etc.	Alternative iron-making processes
ULSAB-AVC	Advanced high-strength steel for automobiles
Electronics industry	
IBM	Energy efficiency in data centres
Yokogawa Electric	Energy-saving controller for air conditioning water pumps
Sharp	Enhancing recycling of electronic appliances
Xerox	Managed print services

The automotive and transport industry is taking steps to reduce GHG emissions and other environmental impacts, notably those associated with fossil fuel combustion. Combined with the growing demand for mobility, particularly in developing economies, many eco-innovation initiatives have focused on increasing the overall energy efficiency of automobiles and transport, while heightening automobile safety. Eco-innovations have, for the most part, been realised through technological advances, typically in the form of product or process modification and redesign, such as more efficient fuel injection technologies, better power management systems, energy-saving tyres and optimisation of painting processes. Yet, there are indications that the understanding of eco-innovation in this sector is broadening. Alternative business models and modes of transport such as the bicycle-sharing scheme in Paris, France (Box 2.1) are being explored, as are new ways of dealing with pollutants from manufacturing processes of automobiles.

The iron and steel industry has in recent years substantially increased its environmental performance through a number of energy-saving modifications and the redesign of various production processes. These have often been driven by strong external pressures to reduce pollution and by increases in the prices and scarcity of raw materials. While most of the industry's eco-innovative initiatives have focused on technological product and process advances, the industry's engagement in various institutional arrangements has laid the foundation for many of these developments. For example, the development of advanced high-strength steel was made possible through an international collaborative arrangement between vehicle designers and steel makers and enabled the production of stronger steel for the manufacturing of lighter and more energy-efficient automobiles (Box 2.2).

The electronics industry has so far mostly been concerned with eco-innovation in terms of the energy consumption of its products. However, as consumption of electronic equipment continues to grow, companies are also seeking more efficient ways to deal with the disposal of their products. As seen in the other two sectors, most eco-innovations in this industry have focused on technological advances in the form of product or process modification and redesign. Similarly, developments in these areas have been built upon eco-innovative organisational and institutional arrangements (see Box 2.3). Some of these arrangements have also, perhaps unsurprisingly, been among the most innovative and forward-looking. A notable example is the use of large-scale Internet discussion groups, dubbed 'innovation jams' by IBM, to harness the innovative ideas and knowledge of thousands of people. Alternative business models, such as product-service solutions instead of merely selling physical products, have also been applied, as exemplified by new services in the form of energy

BOX 2.1 *VÉLIB'*: SELF-SERVICE BICYCLE-SHARING SYSTEM IN PARIS

In an attempt to reduce traffic congestion and improve air quality, the City of Paris introduced a self-service bicycle-sharing system *Vélib'* in the summer of 2007. This system provides 24 000 bicycles in 1750 stations, located every 300 metres throughout the city. Each station is equipped with an automatic rental terminal at which people can hire a bicycle with a small subscription fee that can be linked to the card used for the city's metro and bus system. A subscription allows the user to pick up a bicycle from any station and use it at no charge for 30 minutes. After that a charge is incurred for additional time. The payment scheme was designed to keep bicycles in constant circulation and increase intensity of use. Real-time data on bicycle availability at every station is provided through the Internet and via mobile phones.

By the end of 2009, the number of annual subscribers reached 150 000 and between 65 000 and 150 000 trips are being made each day, making cycling fashionable for Parisians as well as tourists. Part of the success owes to the system's design, with its strong focus on flexibility, availability and ease of use. Its start-up financing, as well as operation for 10 years and associated costs, was undertaken by JC Decaux advertising company, while the City of Paris in return transferred full control of a substantial portion of the city's advertising billboards to this company (OECD, 2010a). Building on this success, this bike-sharing system has been exported to many countries and a new *Autolib'* car-sharing scheme with 3000 electric vehicles was launched in December 2011.

BOX 2.2   THE DEVELOPMENT OF ADVANCED
HIGH-STRENGTH STEEL FOR
AUTOMOBILES

The introduction of new legislative requirements for motor vehicle emissions in the United States in 1993 intensified pressures on the automotive industry to reduce the environmental impact from the use of automobiles. In response, a number of steelmakers from around the world joined together to create the Ultra-Light Steel Auto Body (ULSAB) initiative to develop stronger and lighter autobodies. From this venture, the ULSAB Advanced Vehicles Concept (ULSAB-AVC) emerged. The first proof-of-concept project for applying advanced high-strength steel (AHSS) to automobiles was conducted in 1999.

By optimising the car body with AHSS at little additional cost compared to conventional steel, the overall weight saving could reach nearly 9 per cent of the total weight of a typical five-passenger family car. It is estimated that for every 10 per cent reduction in vehicle weight, the fuel economy is improved by 1.9–8.2 per cent (World Steel Association, 2008). At the same time, the reduced weight makes it possible to downsize the vehicle's power train without any loss in performance, thus leading to additional fuel savings. Owing to their high- and ultra-high-strength steel components, such vehicles rank high in terms of crash safety and require less steel for construction.

management in data centres (IBM) and optimisation of printing and copying infrastructures (Xerox).

To sum up, the primary focus of current eco-innovation in manufacturing industries tends to rely on technological advances, typically with products or processes as eco-innovation targets, and with modification or redesign as principal mechanisms (Figure 2.5). Nevertheless, a number of complementary changes have functioned as key drivers for these developments. In many of the examples, the changes have been either organisational or institutional in nature, such as the establishment of separate environmental divisions for improving environmental performance and directing R&D, or the setting up of intersectoral or multi-stakeholder collaborative research networks. Some industry players have also started exploring more systemic eco-innovation through new business models and alternative modes of provision.

BOX 2.3   ENERGY-SAVING CONTROLLER FOR AIR CONDITIONING WATER PUMPS

Air conditioners function by driving hot or cold water through piping to units located on each level of the building. The amount of cold water varies according to the desired temperature relative to the outside temperature. However, conventional air conditioners operate at the pressure required for maximum heating and cooling demands. Based on research revealing that in Japan air conditioning consumes half of a building's total energy, Yokogawa Electric, a Japanese manufacturer, sought to create a simple, inexpensive and low-risk control mechanism that would eliminate wasteful use of energy. The resulting product, Econo-Pilot, can control the pumping pressure of air conditioning systems in a sophisticated way and can reduce annual pump power consumption by up to 90 per cent. It can be installed easily and inexpensively, precluding the need to buy new cooling equipment. The technology has been successfully applied in equipment factories, hospitals, hotels, supermarkets and office buildings.

Econo-Pilot is based on the technology devised by Yokogawa jointly with Asahi Industries Co. and First Energy Service Company. It was developed and demonstrated through a joint research project with the New Energy and Industrial Technology Development Organization (NEDO), a public organisation established by the Japanese government to coordinate R&D activities of industry, academia and the government.

Image: Yokogawa Electric Corporation.

The heart of an eco-innovation cannot necessarily be represented adequately by a single set of target and mechanism characteristics. Instead, eco-innovation seems best examined and developed using an array of characteristics ranging from modifications to creations across products,

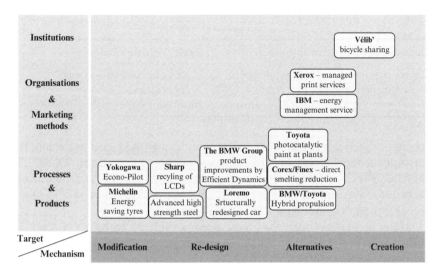

Note: This map only indicates primary targets and mechanisms that facilitated the listed eco-innovation examples. Each example also involved other innovation processes with different targets and mechanisms.

Figure 2.5 Mapping primary focuses of eco-innovation examples

processes, organisations and institutions. The characteristics of a particular eco-innovation furthermore depend on the observer's perspective. The analytical framework can be considered a first step towards more systematic analysis of eco-innovation.[6]

SYSTEMIC CHANGES AND BUSINESS OPPORTUNITIES

Certainly, the scale of improvement necessary to realise green growth is daunting. Given the current status of decoupling as observed above, it has to be clear to government and industry alike that incremental improvement in resource and energy efficiency is not enough to fulfil their long-term commitment. On the other hand, many promising ways to enable more radical improvement in efficiency through further innovation have not been fully exploited.

As the above framework of eco-innovation has shown, first of all, it should be clearly recognised that innovation is not just about developing new technologies but there are diverse approaches that help realise resource efficiency and green growth, including non-technological

Table 2.2   Application of technologies in different types of innovation

Incremental/disruptive innovation	Radical/systemic innovation
Existing (but improved) technologies in existing application	Existing technologies in new application
New technologies in existing application	New technologies in new application

changes. Such approaches can be roughly categorised into incremental, disruptive, radical and systemic innovation, as shown above. Table 2.2 highlights the basic distinction (though there is no clear line to distinguish) between categories based on how technologies – both existing and new, breakthrough technologies – are applied in society.

Many eco-innovations that have been considered to help greening in fact remain incremental or at best disruptive. For example, many efforts to improve the fuel efficiency of vehicles – the development and diffusion of hybrid, electric and fuel-cell cars – are not primarily aimed at changing the nature of mobility based on individual ownership and road infrastructure, and therefore they do not help tackle the continuous rise of congestion and vehicle ownership, particularly in emerging economies, which lead to the overall increase in environmental impacts. Incremental and disruptive innovation, as the dominant form of innovation in the marketplace, has been helping a relative decoupling of growth and environmental pressures and should still continue to be a very important part of greening efforts, but it alone will not be sufficient to cope with the challenges the world is facing today.

The past success of the radical reduction and absolute decoupling of emissions of harmful substances such as SOx, NOx and chlorofluorocarbons (CFCs) in fact owes much to the discovery and application of safer alternative substances or production processes, not incremental improvement. Even though many barriers need to be overcome to implement it, as discussed below, more radical and systemic innovation should therefore be essential to enable the absolute decoupling of other environmental impacts from economic growth.

One of the areas where clear benefits of more systemic innovation have been well exemplified is general-purpose technologies. While the information and communication technologies (ICTs) urgently need to raise energy efficiency in existing products which are responsible for around 2 per cent of global GHG emissions, one estimate indicates that the transformation of the way people live and businesses operate through the smart application of ICTs could reduce global emissions by 15 per cent by 2015 (Climate Group, 2008). If safely applied, biotechnology and nanotechnology could also create substantial environmental benefits, mainly through unique applications in different sectors and convergence with other technologies (OECD, 2009b).

Even more interestingly, pioneering businesses started looking into new ways of creating value from sustainability, particularly innovation in business models of delivering services to end users. The small sample of manufacturing firm activities reviewed in this chapter indicates that although the primary focus of the firm's activities to enhance sustainability tends to be technological developments and advances with products or processes, understanding of green innovation is also broadening as alternative business models and new modes of provision are being explored, particularly by new firms and public–private partnerships (OECD, 2010b). A Green Paper to the Nordic Council of Ministers (FORA, 2010) reports the results from case studies of 'green business models' in the Nordic countries, which include five types of product-service systems (PSSs)[7] (Box 2.4). The study concludes that many such business models have the potential to generate solid business cases and jobs, while leading to significantly lower environmental impacts and supporting the transition towards green growth.

To look into how business opportunities will be developed in the long-term, the World Business Council for Sustainable Development (WBCSD) developed the Vision 2050 jointly with member multinational companies. This vision shows that the expected economic transformations represent opportunities in a broad range of business segments as they see the challenges of growth, urbanisation, resource scarcity and environmental change to become key strategic drivers for business in the coming decades. Opportunities range from developing and maintaining low-carbon, zero-waste cities and infrastructure, to improving and managing ecosystems and lifestyles. Enabling these changes is also considered to be creating opportunities for the finance and ICT sectors.

CONCEPTUALISING BUSINESS MODELS FOR GREEN GROWTH

Overall, there are a wide range of economic opportunities for leveraging on eco-innovation by placing it at the core of business strategies. To capture such future opportunities, make them into a commercial success and disseminate good practices, both industry and policy-makers need to better understand the social, technical and political factors enabling or obstructing such eco-innovation (Figure 2.6). Among the key elements in determining the success of eco-innovation, a special focus needs to be

---

BOX 2.4   EXAMPLES OF 'GREEN BUSINESS MODELS'

A paper to the Nordic Council of Ministers defines 'green business models' as 'business models which support the development of products and services (systems) with environmental benefits, reduce resource use and waste and which are economic viable'. The following five types of business models were distinguished in the study:

Functional sales: The provider offers the customer to pay for the functionality or result of the product instead of buying the product itself. The structure of the business model gives the provider the incentives to optimise and maintain the product to ensure lifecycle cost effectiveness and reduce the overall environmental impact.

Energy saving company (ESCO): This type of firm provides companies and public buildings with energy-saving solutions such as the installation of combined heat and power (CHP) equipment and in return is paid by part of the savings achieved, not by the equipment. This encourages the diffusion of large energy-saving equipment as the customer does not have to pay all the cost up front.

Chemical management services (CMS): This type of firm engages in a strategic, long-term contract to supply and manage the customer's chemicals and related services. The provider of CMS is typically remunerated in some form of the customer's output (e.g. painted car doors).This gives the provider the incentives to reduce the use of chemicals.

Design, Build, Finance and Operate (DBFO): This business model involves long-term contracts over the construction, maintenance and operation phase (typically 20–30 years) of projects such as roads, buildings and facilities. This gives incentives to improving the quality of the construction project so that the lifecycle costs would be lowered.

Sharing: Instead of private ownership, the product is shared among a number of users when the individual users need access to the product. The economic benefits of this model are less evident than in the other business models, but the sharing of products may pave the way for new products to the market.

Source: FORA (2010).

---

put on the business model, which brings out eco-innovation to the market and promotes its deployment and diffusion. According to Osterwalder et al. (2010: 14), 'a business model describes the rationale of how an organization creates, delivers, and captures value'. The business model is also

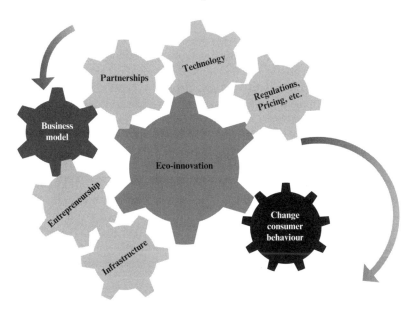

Figure 2.6 Various factors surrounding eco-innovation

understood as a holistic approach towards explaining how firms conduct business (Zott et al., 2010).

The business model approach offers a comprehensive way to understand how value is created and distributed. Eco-innovation aims to create both economic and environmental value, while a business model acts as a value driver and enabler of green technologies and solutions. The focus on business models allows for a better understanding on how environmental value is captured, turned into profitable products and services, and can deliver convenience and satisfaction to users. In concrete terms, the analysis of eco-innovation cases sheds light on whether, to what extent and how environmental values are reflected in their value propositions, customer segmentation, use of resources and collaboration patterns as well as managing cost and revenue streams.

By replacing old business practices, innovative business models could also allow firms to restructure their value chain and generate new types of producer–consumer relationships, and alter the consumption culture and use practices. The business model perspective is therefore particularly relevant to radical and systemic eco-innovation. With the major challenges for green growth in mind, it should also be investigated how business models and strategies can induce and help diffuse more radical eco-innovation and enable systemic changes and transformation, along with investigating the business models of eco-innovation cases themselves.

Business models combine all the core components of business strategies and operations that create and deliver value to the customers as well as to the firm. The components of business models typically include strategic decisions on customer segmentation, products and services (or value propositions) to offer, business and research partners to engage with, resources to create and channels to deliver value, as well as the underlying cost structure and revenue streams to ensure economic viability of business.

In the typical approaches to business models, however, environmental sustainability is rarely at the core of value propositions. Although the business community increasingly recognises the challenges of climate change and resource scarcity, these issues are not automatically internalised in the building blocks of the firm's strategy and operations. Nevertheless, the business model concept has recently been adopted in the discourse on sustainable services (Halme et al., 2007) and is also routinely utilised in studies on PSSs.

When looking at eco-innovation cases, their business models go through different degrees of change. Most changes seem to take place in the activities component with research and development (R&D) and product and process development. In the service-oriented models such as functional sales and car sharing, the change is expressed in the overall shift from product to service provision. Another most observed shift is associated with reconfiguring the relationship with conventional customers or building relationships with new customers (or both). New markets and customers are targeted in such models as waste regeneration (new farmers/users of soil conditioner), ICT solutions (larger coverage of people, firms, utility companies), energy service companies (ESCOs) (new companies willing to cut energy consumption), and buyers of renewable energy. Approaching new customers or changing relationships with conventional customers often also requires the transformation of channels to customers. The business model applied in waste regeneration systems seem to go through substantial transformation, outreaching all components. Multi-actor business models of industrial symbiosis and eco-cities also appear to be about changing many components, while keeping themselves in the same market niche and servicing the same customers (Table 2.3).

Meanwhile, it should be noted that the environmental benefits associated with certain business models depend greatly on the way the products are used by customers. For instance, sharing of products may entail negative environmental impacts, if the access to a shared product increases the customers' use of the product (e.g. car sharing may increase the total use of car mobility by improving access to those who otherwise do not use cars)

Table 2.3   Business model eco-innovation: change in elements

Eco-innovation/ business model types	Value proposition	Business operations			Customer aspects		
		Key activities	Key partners	Key resources	Customer segments	Customers relations	Customer channels
Green value-added products	*Products with better performance, savings*	R&D	Changes of suppliers (not always)	Other resources	New customers/ market		
Renewable energy-based systems	*Cheaper and cleaner energy*	R&D			New customers/ market	New relationship	New relationship
Efficiency optimisation by ICT	*Economic savings due to more efficient management of resources*				New customers/ market	New relationship	
Functional sales	*More efficient services*	R&D (not always)			New customers/ market	New relationships	
Innovative financing	*Resource saving*	Shifting to new services				New relationships	New channels
Sustainable mobility systems	*Flexibility, savings for customers*	Shifting to new services	New partners			New relationships	New channels
Industrial symbiosis	*Resource saving, higher efficiencies*	R&D	Reconfigured network of partners	New expertise		New relationship	New relationship
Eco-cities	*Improved life quality, convenience*	Development	New network of partners	New expertise		New relationship	New relationship

Source: Technopolis Belgium.

43

or if the environmental impact from logistics needed to send and collect the shared items exceeds the benefits gained from product sharing (FORA, 2010; Tukker and Tischner, 2004; EPA, 2011).

POLICIES FOR ENABLING ECO-INNOVATION AND GREEN GROWTH

Needless to say, there are many barriers to enabling the diverse types of eco-innovation described so far. Policy-makers and industry are increasingly facing difficulties in investing in long-term future due to short political cycles and pressure from shareholders. Sector or technology-based approaches in conventional environmental policies may fail to take into account the full innovation cycle of environmental technologies and undermine opportunities for cross-sectoral application of new technologies. The market-based 'getting prices right' measures such as carbon taxes and emissions trading schemes may not be enough to guide investment in promising technologies with high initial cost and much-needed green infrastructures.

One of the most effective ways to break the barriers for eco-innovation is an increasing understanding and utilisation of 'demand-pull' levers in policies and business development. Policy-makers, industry and investors alike – even also many green theorists – have tended to focus their efforts on the earlier stages of innovation such as R&D and technological sophistication, that is, 'supply-push' factors, as well as the effectiveness of economic instruments for improving the general environment for innovation. They have not fully taken into account the fact that many promising eco-innovations have to walk through a 'valley of death' in the face of commercialisation, and the purchasing decisions made by users and consumers have major impacts on the extent to which markets can make eco-innovations successful.

Policies that could drive demand and change consumer behaviour to help introduce and diffuse eco-innovation include 'smart regulations' oriented towards the delivery of performance and outcomes (vis-à-vis command-and-control regulations); technical standardisation (e.g. ISO standards); price incentives and disincentives (subsidies and penalties to final users); labelling and certification that help consumer choice; and public procurement (Figure 2.7). Denmark's experience with feed-in-tariffs in stimulating the wind power industry between the mid-1980s and the late 1990s is often cited as a good example of demand-side policies. France's bonus-malus programme which was introduced for personal cars in 2007 is unique in its 'carrot and stick' approach. This policy provides

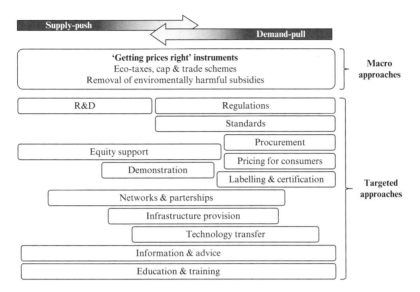

Figure 2.7   Overview of policy instruments for eco-innovation

rewards to consumers buying cars with low CO_2 emissions and penalties for those buying cars with high emissions. The results to date show a clear shift in consumer choices. Over half of newly registered cars in 2009 were those with low CO_2 emissions (less than 130g/km), while those with high emissions (over 160g/km) accounted for less than 10 per cent of the newly registered cars (OECD, 2011b).

Many radical and systemic solutions have, however, been facing rather high entry barriers since they do not necessarily fit the existing technology systems, let alone the current patterns of production and lifestyles, and need long-term investment in the development of new infrastructures as well as cultural changes. To break such technological lock-in and path-dependencies, policy decisions taken today need to incorporate a longer time horizon and government needs to take stronger leadership in actively supporting infrastructure and platform development that leverages the radical reduction of environmental impact, for example in the development of smart grids, public transport systems and green cities.

Among different infrastructures that enable radical and systemic eco-innovation, the role of ICTs has been emphasised as they would allow inconveniences for customers to be minimised and the efficiency of the system to be maximised. For example, the software used in electric vehicles, which was designed and developed by Israel's Better Place, provides the driver with complete information by displaying the energy level in the battery, locating the nearest battery recharging and swapping facilities, and allowing the driver to handle their booking, parking and charging spots conveniently (Meenakshisundaram and Shankar, 2010). Similarly, many service-oriented business models such as ESCO and video-conferencing services are very much dependent on ICTs which enable monitoring and control of data and information (EPA, 2011).

The availability of supporting physical infrastructures is also often an important factor for eco-innovative solutions. The success of the biogas-based transport system in Linköping, Sweden was to a large degree owed to the specifically designed refuelling stations for public buses and private cars (Martin, 2009). The Better Place electric car-sharing system developed its automated battery swapping facilities which can replace depleted batteries with charged batteries within three minutes without drivers getting out of the car (Meenakshisundaram and Shankar, 2010).

To set priorities for investment in 'green infrastructures', a new mechanism needs to be developed to identify where technologies can have the highest environmental, economic and social benefits, while international collaboration and public–private partnerships need to be encouraged to share costs and risks and ensure wider diffusion. The use of scenario studies, technology foresight or roadmaps can provide insights on the scope for technological progress and innovation in different areas and may therefore help in guiding more collective, deliberative policy decisions (OECD, 2011b).

CONCLUSIONS: TOWARDS A COMPREHENSIVE ECO-INNOVATION ANALYSIS

In order to meet global environmental challenges such as climate change, much attention has been paid to innovation as a way to develop sustainable solutions. This chapter has aimed to draw the reader's attention to many innovation possibilities to make decoupling and green growth possible, and to elaborate the understanding of innovation processes. From the perspective of eco-innovation, the primary focus of sustainable production practices tends to be on technological advances for the modification and redesign of products or processes. However, some advanced industry players have adopted complementary organisational or institutional changes, for example, synergising resource flows among different industries and offering product-service solutions rather than selling physical products.

As such, it is essential to capture both incremental and systemic types of eco-innovation, unlike the conventional economic and empirical research in this area. The former type of innovation mainly supports realising relative decoupling in the relatively short term, while the latter has potential for enabling absolute decoupling in the long term. Although improvements in eco-efficiency through incremental innovations have led to substantial environmental progress, the gains have often been offset by increasing consumption or outpaced by scale effects. Countries will therefore need to engage in a broader range of eco-innovations.

Probably most needed for government is knowledge and competence to set balanced priorities between taking short-term 'low-hanging fruit' and investing in long-term substantial changes. The potential economic and environmental benefits of systemic innovation need to be identified, particularly where applications of new technologies can have the greatest benefits. To guide the processes of system transition and industry restructuring, visions and scenarios for future societal systems should be collectively developed and shared in different areas such as transport, housing and nutrition.

Among the key social, technical and political elements in determining the success of eco-innovation, particular attention needs to be paid to the business model, which brings out eco-innovation to the market and enables its dissemination. The business model perspective allows for a better understanding of how environmental value is captured, turned into profitable products and services, and can deliver convenience and satisfaction to users. This perspective is particularly relevant to radical and systemic eco-innovation as it also allows us to understand eco-innovation as the way to restructure the value chain and producer–consumer relationships, and to alter consumer practices.

The effort towards a comprehensive understanding of the mechanisms and drivers of eco-innovation and how it enables green growth is still in its infancy, and needs to be quickly accelerated to help governments tackle global warming and help industry cope with or even gain from the anticipated green transition. To contribute to such efforts, the OECD project on Green Growth and Eco-Innovation is further working on case studies of new business approaches to eco-innovation. The analysis focuses on the innovation processes of specific cases collected from member countries, including the source of the original idea, the business model, the role of partnerships and collaboration, the impact of policies in facilitating the innovation, the sources of funding and the potential economic and environmental benefits.

ACKNOWLEDGEMENTS

The author is a Senior Programme Officer at the International Renewable Energy Agency (IRENA) and a former Senior Policy Analyst at the

Organisation for Economic Co-operation and Development (OECD). This chapter is written based on the progress of the OECD project on Green Growth and Eco-Innovation. The author would like to acknowledge the financial support for this project from the European Commission DG Environment, and the substantive contributions by Technopolis Belgium sprl (Asel Doranova, Liina Joller, Lorena Rivera Leon, Michal Miedzinski and Geert van der Veen) and Karsten Bjerring Olsen to this project. The views expressed in this chapter do not represent the views of the OECD or the IRENA.

NOTES

1. For the latest developments in the OECD Green Growth Strategy, see www.oecd.org/greengrowth.
2. For more details on this OECD project, see www.oecd.org/innovation/green.
3. In December 2011, the EU renewed the Environmental Technology Action Plan (ETAP) as the Eco-Innovation Action Plan. The new plan reflects the extension of the eco-innovation concept by embracing non-technological aspects of eco-innovation such as innovation in business models and increasing attention to the diffusion and commercialisation stages of eco-innovation on top of the conventional focus on research and development (R&D).
4. For more details go to the EIO website, www.eco-innovation.eu.
5. A few notable examples are illustrated in boxes. For detailed information on each example, see OECD (2010).
6. A combination of this eco-innovation framework with the frameworks of system transition developed by some scholars (e.g. Geels, 2005; Loorbach, 2007; Carrillo-Hermosilla et al., 2009) could further help understand the dynamic nature of radical changes created by eco-innovations.
7. A PSS is a business model that has 'tangible products and intangible services designed and combined so that they jointly are capable of fulfilling specific customer needs' (Tukker, 2004).

REFERENCES

Carrillo-Hermosilla, J., P. del Río Gonzaléz and T. Könnölä (2009), *Eco-Innovation: When Sustainability and Competitiveness Shake Hands*, New York: Palgrave Macmillan.
Environmental Protection Agency (EPA), United States (2011), 'Green servicizing: building a more sustainable economy', internal working draft, Washington, DC: EPA.
Geels, F.W. (2005), *Technological Transitions and System Innovations: A Co-Evolutionary and Socio-Technical Analysis*, Cheltenham, UK and Northampton, MA, USA: Edward Elgar.
Halme, M., Markku Anttonen, Mika Kuisma, Nea Kontoniemi and Erja Heino (2007), 'Business models for material efficiency services: conceptualization and application', *Ecological Economics*, **63**(1): 126–37.

Hellström, T. (2007), 'Dimensions of environmentally sustainable innovation: the structure of eco-innovation concepts', *Sustainable Development*, **15**(3): 148–59.

Loorbach, D. (2007), *Transition Management: New Mode of Governance for Sustainable Development*, Utrecht: International Books.

Martin, M. (2009), 'The "biogasification" of Linköping: a large technical systems perspective', *Environmental Technology and Management*, Linköping: Linköpings Universitet.

OECD (2007), *Science, Technology and Industry Scoreboard 2007: Innovation, and Performance in the Global Economy*, Paris: OECD.

OECD (2008), *Open Innovation in Global Networks*, Paris: OECD Publishing.

OECD (2010a), *Eco-Innovation in Industry: Enabling Green Growth*, Paris: OECD Publishing.

OECD (2010b), *The Impacts of Nanotechnology on Companies: Policy Insights from Case Studies*, Paris: OECD Publishing.

OECD (2011a), *Towards Green Growth*, Paris: OECD Publishing.

OECD (2011b), *Fostering Innovation for Green Growth*, Paris: OECD Publishing.

Porter, E.M. and C. van der Linde (1995), 'Towards a new conception of the environment–competitiveness relationship', *Journal of Economic Perspectives*, **9**: 97–118.

Rennings, K. (2000), 'Redefining innovation: eco-innovation research and the contribution from ecological economics', *Journal of Ecological Economics*, **32**: 319–32.

Scrase I., Andy Stirling, Frank Geels, Adrian Smith and Patrick Van Zwanenberg (2009), 'Transformative innovation: a report to the Department for Environment, Food and Rural Affairs', Science and Technology Policy Research (SPRU), University of Sussex, Brighton.

Smith, K. (2008), 'The challenge of environmental technology: promoting radical innovation in conditions of lock-in', final report to the Garnaut Climate Change Review, Australian Innovation Research Centre.

Steward, F. (2008), 'Breaking the boundaries: transformative innovation for the global good', NESTA provocation 07, April, London.

Tukker, A. (2004), 'Eight types of product-service systems: eight ways to sustainability?' *Business Strategy and the Environment*, **13**: 246–60.

Tukker, A. and U. Tischner (2006), *New Business for Old Europe: Product-service development, competitiveness and sustainability*, Sheffield: Greenleaf Publishing.

World Business Council for Sustainable Development (WBCSD) (2010), *Vision 2050*, Geneva: WBCSD.

Zott, C., R. Amit and L. Massa (2010), 'The business model: theoretical roots, recent development, and future research', IESE Business School Working Paper, No. 862, University of Navarra, Pamplona.

Web References

Charter, M. and T. Clark (2007), 'Sustainable innovation: key conclusions from sustainable innovation conferences 2003–2006', Farnham: Centre for Sustainable Design, available at http://cfsd.org.uk/Sustainable%20Innovation/ Sustainable_Innovation_report.pdf (accessed 14 June 2012).

Climate Group (2008), 'SMART 2020: enabling the low carbon economy in the information age', report on behalf of the Global e-Sustainability Initiative

(GeSI), London: Climate Group, available at www.smart2020.org/_assets/ files/02_Smart2020Report.pdf (accessed 14 June 2012).

FORA (2010), 'Green business models in the Nordic region: a key to promote sustainable growth', Green Paper for the Nordic Council of Ministers, FORA, Copenhagen, available at www.foranet.dk/media/27577/greenpaper_ fora_211010.pdf (accessed 14 June 2012).

Future Think (2008), 'Future of green business strategy', *Futurist Report*, available at http://futurethinktank.com/2008/08/21/the-future-of-green-business-strategy (accessed 14 June 2012).

International Energy Agency (IEA) (2010), *Energy Technology Perspectives 2010: Scenarios and Strategies to 2050*, Paris: OECD/IEA, available at www.iea.org/ techno/etp (accessed 14 June 2012).

Kemp, René and Peter Pearson (2008), 'MEI project about Measuring Eco-Innovation: final report', under the EU 6th Framework Programme, Maastricht: MERIT, available at www.merit.unu.edu/MEI/deliverables/MEI%20D15%20 Final%20report%20about%20measuring%20eco-innovation.pdf (accessed 14 June 2012).

Meenakshisundaram, R. and B. Shankar (2010), 'Business model innovation by better place: a green ecosystem for the mass adoption of electric cars', Oikos sustainability case collection, ICMR Center for Management Research, Hyderabad, available at www.oikos-international.org/fileadmin/oikos-international/interna tional/Case_competition/Competition_2010/CS_Track/oikos_CWC_CS_TRA CK_2010_3rd_Place_Better_Place.pdf (accessed 14 June 2012).

Ministry of Economy, Trade and Industry, Japan (METI) (2007), 'The key to innovation creation and the promotion of eco-innovation', report by the Industrial Science Technology Policy Committee, Tokyo: METI (Japanese only), available at http://warp.da.ndl.go.jp/info:ndljp/pid/281883/www.meti. go.jp/press/20070706003/20070706003.html (accessed 14 June 2012).

OECD (2009a), *Declaration on Green Growth*, adopted at the Council Meeting at Ministerial Level, 25 June, Paris: OECD, available at www.oecd.org/document/ 63/0,3343,en_2649_201185_43164671_1_1_1_1,00.html (accessed 14 June 2012).

OECD (2009b), 'Sustainable manufacturing and eco-innovation: towards a green economy', OECD Policy Brief, June, Paris: OECD, available at www.oecd.org/ dataoecd/34/27/42944011.pdf (accessed 14 June 2012).

OECD and Statistical Office of the European Communities (Eurostat) (2005), *Oslo Manual: Guidelines for Collecting and Interpreting Innovation Data*, 3rd edn, Paris: OECD, available at www.oecd.org/document/33/0,3343, en_2649_34409_35595607_1_1_1_1,00.html (accessed 14 June 2012).

Osterwalder, A., Y. Pigneur and A. Smith (2010), 'Business Model Generation', available at *www.businessmodelgeneration.com* (accessed 14 June 2012).

United Nations Environment Programme (UNEP) (2011), 'Towards a green economy: pathways to sustainable development and poverty eradication', Nairobi: UNEP, available at www.unep.org/greeneconomy/Portals/88/docu ments/ger/ger_final_dec_2011/Green%20EconomyReport_Final_Dec2011.pdf (accessed 14 June 2012).

3. Sustainable development through innovation? A social challenge

Corinne Gendron

INTRODUCTION

Innovation is valued in our societies, and people have faith in its ability to solve human problems, even if the past shows a mixed record. This is why we should not be surprised that, as for other important human challenges, people and institutions turn to technology as a possible solution to the environmental crisis, forgetting that science and technology are also at the origin of this particular crisis.

But the innovation challenge is now better understood: after having been presented as a miraculous process that would naturally solve the ecological crisis, it is more and more seen as a tool that needs to be managed in order to accomplish what is needed.

In this chapter, I propose to analyze how sustainable development challenges have become more precise and might lead to a new generation of public policies regarding the economy and innovation. Those policies would adopt a new position towards the market and the economy, recognizing that they have to be regulated differently to produce wealth instead of negative externalities. Moreover, they would more actively engage private actors in sustainability innovation processes, recognizing that new actors, like civil society, can also play a role. And finally, innovation has to be managed in order to respond to society's challenge of sustainable development.

FROM SUSTAINABLE DEVELOPMENT TO GREEN ECONOMY

A lot has been said about sustainable development, but conceptual confusion remains as it is often reduced to environmental considerations within corporate strategies, or understood as sustainable growth in the economic sphere. This confusion is not surprising, given that as the concept took on political dimensions, its imprecision was a guarantee for its successful dissemination in research and wide acceptance by political actors. Daly (1990) even argues that the consensus reached around the concept has been at the cost of intrinsic contradictions in the Brundtland report:

> The Brundtland Commission Report has made a great contribution by emphasizing the importance of sustainable development and in effect forcing it to the top of the agenda of the United Nations and the multilateral development banks. To achieve this remarkable consensus, the Commission had to be less than rigorous in avoiding self contradiction. (Daly, 1990: 1)

What was at stake in the Brundtland report, and the Rio conference that followed, was the apparent contradiction between environmental limitations and social aspirations to development, especially for the developing nations. As environmental degradation was recognized and protection of the environment was institutionalized through new ministries in developed countries, at the international level the necessity for ecological preservation was highlighted. But environmental protection was a potential threat to the 'right to development' of the poorest countries. Moreover, it was obvious that to protect their resources and the global environment, developing countries would need technical and financial support from the North. This was clearly stated 20 years before Rio Earth Summit, in the Stockholm Declaration on the Human Environment (UN, 1972: 2):

> Environmental deficiencies generated by the conditions of underdevelopment and natural disasters pose grave problems and can best be remedied by accelerated development through the transfer of substantial quantities of financial and technological assistance to supplement the domestic efforts of the developing countries, and such timely assistance as may be required (principle 9).

The idea of sustainable development, which links development goals to environmental protection, can thus be seen as a positive result of developing countries' claims for their development. But this tricky union did not resolve the problematic articulation of economic productivism and environmental conservation, and some even say that it was misleading in proposing their possible reconciliation. Presented as it was, it could imply that development was a condition for environmental protection, while recognizing that environmental degradation will worsen underdevelopment. From an environmental point of view, this perspective might lead to a dead end since development relies on a complex geopolitical context that enhances its uncertainty; as uncertain as it is, to present development as a condition might annihilate any attempt to protect the environment. But it also hides what might be the real question at stake: which development are we talking about?

During recent decades, even if the conception of development has radically changed from its early theorizations by Rostow (1960), it has remained correlated with the necessity for economic growth. Economic growth is still presented as a necessary condition of development, and therefore stands as a goal for public policies. The problem is that the content, or what we might call the quality of this growth which determines its ecological intensity and social outcomes was not questioned, in the South as well as in the North. Therefore, policies aiming at development as economic growth by any means could easily be in contradiction with environmental protection objectives. Indeed by simply fostering development that would take into account environmental and social as well as economic considerations, the sustainable development concept did not provide a clear path to reconcile the economy with environmental conservation. One could argue, for example, that it suffices to stimulate the traditional economy while creating conservation areas in order to comply with a sustainable development engagement.

This is radically changing with the new Green Economy concept proposed by a recent United Nations Environment Programme (UNEP) report and discussed at Rio+20. Entitled *Towards a Green Economy: Pathways to Sustainable Development and Poverty Eradication – A Synthesis for Policy Makers* (UNEP, 2011), the report insists on the necessary transition of the economy to face the challenge of environmental issues, and more specifically of climate change. The definition of the green economy puts an emphasis on carbon emissions, preservation of resources and social inclusion, and states that growth must be driven by public and private investments directed toward the reduction of society's environmental pressures:

> [UNEP defines a green economy as one that results in] *improved human well-being and social equity, while significantly reducing environmental risks and ecological scarcities.* In its simplest expression, a green economy can be thought of as one which is low carbon, resource efficient and socially inclusive. In a green economy, growth in income and employment should be driven by public and private investments that reduce carbon emissions and pollution, enhance energy and resource efficiency, and prevent the loss of biodiversity and ecosystem services. These investments need to be catalysed and supported by targeted public expenditure, policy reforms and regulation changes. The development path should maintain, enhance and, where necessary, rebuild natural capital as a critical economic asset and as a source of public benefits, especially for poor people whose livelihoods and security depend on nature. (UNEP, 2011: 3)

The social is presented here as a goal of this new economy, in a vision that rejects the intrinsic contradiction between social equity and development on one side, and environmental conservation on the other, but confirms at the same time the necessity of a new economic model. The role of the state in this economic transition is central, leading toward what can be called a green interpretation of Keynes's interventionism, with an explicit reference to the post-Great Depression model: a Green New Deal. Enabling conditions to the green economy include, among others, sound regulatory frameworks, eco-conditionality of government spending, taxes and market-based instruments. These propositions originated from the UN's initiative following the 2008 financial and economic crisis.

UNEP's work on the green economy raised the visibility of this concept in 2008, particularly through the call for a Global Green New Deal (GGND). The GGND recommended a package of public investments and complementary policy and pricing reforms aimed at kick-starting a transition to a green economy while reinvigorating economies and jobs and addressing persistent poverty. Designed as a timely and appropriate policy response to the economic crisis, the GGND proposal was an early output from the United Nations Green Economy Initiative. This initiative, coordinated by UNEP, was one of the nine Joint Crisis Initiatives undertaken by the Secretary-General of the UN and his Chief Executives Board in response to the 2008 economic and financial crisis (UNEP, 2011: 3).

Given the necessity of public investment that transcended ideological debates, the 2008 crisis was an exceptional opportunity to accelerate the transition to a green economy; even if the share of climate change and environmental protection measures in total public investments to solve that crisis has been very variable, from 80 percent of total investments in South Korea and 38 percent in China, to 21 percent in France, 12 percent in the United States, and only 8 percent in Canada (HSBC, 2009). It also should be noted that in UNEP's report, this green interventionism is not presented in contradiction to a hypothetic *laissez-faire* policy, but in replacement of today's interventionism towards a 'brown economy'. For example, it firmly denounces subsidies on fossil fuels:

> price and production subsidies for fossil fuels collectively exceeded US$650 billion in 2008, and this high level of subsidization can adversely affect transition to the use of renewable energies. In contrast, enabling conditions for a green economy can pave the way for the success of public and private investment in greening the world's economies. (UNEP, 2011: 2)

What is interesting about the green economy concept is not so much that it can replace a notion that, in the eyes of many, has proven to be inapplicable, but that by building on the new development paradigm proposed by sustainable development, it defines more precisely the role that the economy has to play in it. One of the biggest problems encountered with the sustainable development definition has been the respective role of each of its three pillars: the economy, the social and the environment. Saying that it was necessary to take the three pillars into account, often understood as the search for an equilibrium between the three, often led to the idea that an equivalence or a substitution could be made between the environment and the economy (Daly, 1990). Numerous authors have discussed whether or not this substitution is possible in articles on the issue of weak and strong sustainability (Pearce and Atkinson, 1993: 64; Goodland, 1995: 30; Faucheux, 1995: 64). The concept of green economy presents the challenge differently, by insisting on the status of the economy as a means and not an end in itself; in this perspective, the economy is not seen as a given, but must be adjusted to new conditions related to environmental fragility to attain a goal: human and societal development. This perspective is compatible with a definition my colleague and I proposed in earlier work (Gendron and Reveret, 2000), stating that each pillar has a particular function in sustainable development: economy is a means, ecosystems integrity is a condition, and the social dimension is the objective of sustainable development; whereas equity is a condition, a means and a goal all at the same time (see Figure 3.1).

The green economy concept opens a space to understand the malleability of the economy, specifically that it can be shaped and organized to reduce human impact on the environment, and better fulfill its social goal. In a way, it re-politicizes an economy that has tended to be presented as an autonomous sphere responding to rules of its own, with a positive

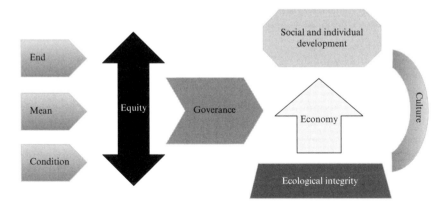

Figure 3.1   A new representation of sustainable development

dynamic that any political intervention would jeopardize. From an ideological point of view, it is interesting to note that this new concept rejects
any positioning in the debate between a directed or a 'free' economy: 'A
green economy does not favour one political perspective over another. It
is relevant to all economies, be they state or more market-led. Neither is
it a replacement for sustainable development' (UNEP, 2011: v). But this
perspective shows how today's economies, far from being autonomous,
are oriented by different policies that foster ecological intensity, and that
this has to change in order to attain a sustainable development model.

> During the last two decades, much capital was poured into property, fossil fuels
> and structured financial assets with embedded derivatives, but relatively little in
> comparison was invested in renewable energy, energy efficiency, public trans-
> portation, sustainable agriculture, ecosystem and biodiversity protection, and
> land and water conservation. Indeed, most economic development and growth
> strategies encouraged rapid accumulation of physical, financial and human
> capital, but at the expense of excessive depletion and degradation of natural
> capital, which includes our endowment of natural resources and ecosystems. By
> depleting the world's stock of natural wealth – often irreversibly – this pattern
> of development and growth has had detrimental impacts on the well-being of
> current generations and presents tremendous risks and challenges for future
> generations. The recent multiple crises are symptomatic of this pattern.
>   Existing policies and market incentives have contributed to this problem
> of capital misallocation because they allow businesses to run up significant
> social and environmental externalities, largely unaccounted for and unchecked.
> Unfettered markets are not meant to solve social problems, so there is a need
> for better public policies, including pricing and regulatory measures, to change
> the perverse market incentives that drive this capital misallocation and ignore
> social and environmental externalities. Increasingly too, the role of appropri-
> ate regulations, policies and public investments as enablers for bringing about
> changes in the pattern of private investment is being recognized and demon-
> strated through success stories from around the world, especially in developing
> countries. (UNEP, 2011: 1–2)

The issue brought up by the idea of the green economy about the ecological and social quality of the economy will raise new debates, forcing
political discourse to explain investment choices on new grounds. To
simply use 'the economy' as a leitmotif to justify a political choice will
be harder, as it will become necessary to explain which kind of economy
public policy is fostering and why. In other words, an economic choice
shall be justified in terms of its ecological impact as well as its maximiza-
tion of social outcomes in comparison with available alternatives. The
path to sustainable development suggested by the green economy will
become a new framework for institutions and organizations; they will
have to integrate social and ecological parameters as never before.

The green economy will not, however, resolve the blurred social dimension that has always been problematic in defining sustainable development. As mentioned above, in the building of the sustainable development concept, the social dimension was introduced as a necessary parameter to be taken into account, but the issue at stake remained environmental deterioration. It was environmental protection that was meant to be adopted into the international agenda with the sustainable development concept. International debates about the environmental crisis enabled the analysis of how it deepens social inequities and redefines social dynamics, but without showing the road to social development. Indeed, social equity and human development have been discussed for centuries, their recognition and management have always been a matter of a political and ideological debate (Rist, 1996) that has not vanished with the sustainable development concept. Neither will they be solved through the concept of the green economy. Even with its reference to poverty eradication, the green economy program proposed by UNEP does not give a clearer view or a precise program for the social sphere of sustainable development, which remains a subject of political and ideological debate. The chapter devoted to poverty eradication insists on how environmental degradation aggravates poverty, while the one devoted to employment and social equity concerns job opportunities in green sectors. Another chapter deals with urban living and low-carbon mobility. As we can see, the report does not really address structural inequity, distribution of wealth or social solidarity and cohesion as central issues.

SUSTAINABLE DEVELOPMENT AND SOCIAL RESPONSIBILITY

The social sphere has been particularly problematic for organizations, where sustainable development is often reduced to its environmental imperatives. Confusion also arose among organizations because the sustainable development concept spread in political discourse at the same time as the social responsibility concept was popularized. Indeed, during the last decade, the two expressions have been used interchangeably by organizations and their definitions have been mixed up. However, with the knowledge acquired on these issues and the experience shared by more and more organizations, they are now starting to be distinguished, leading to a better understanding and allowing useful articulations to be drawn between the two. Released recently, the ISO 26000 standard provides a better vision of this articulation, firstly by defining the two expressions, and secondly by distinguishing between them. To explain sustainable development, it takes up the Bruntland definition, adding the International Union for Conservation of Nature and Natural Resources (IUCN) points about the economic, social and environmental necessary integration. It specifics that: 'The objective of sustainable development is to achieve sustainability for society as a whole and the planet'. The social responsibility definition is far more detailed, as it is the core subject but also the goal of ISO 26 000 standard, the aim of which is to clarify its understanding. It presents its causal foundations, that is, the fact that an organization is responsible for its impacts, and lists its characteristics and constitutive elements.

The ISO definition of sustainable development is:[1]

> **2.1.23 sustainable development** Development that meets the needs of the present without compromising the ability of future generations to meet their own needs NOTE   Sustainable development is about integrating the goals of a high quality of life, health and prosperity with social justice and maintaining the earth's capacity to support life in all its diversity. These social, economic and environmental goals are interdependent and mutually reinforcing. Sustainable development can be treated as a way of expressing the broader expectations of society as a whole. (ISO 26 000, 2010)

The ISO definition of social responsibility is:

> **2.1.18 social responsibility** Responsibility of an **organization** (2.1.12) for the impacts of its decisions and activities on society and the **environment** (2.1.5), through transparent and **ethical behaviour** (2.1.6) that
>
> – contributes to **sustainable development** (2.1.23), including health and the welfare of society;
> – takes into account the expectations of **stakeholders** (2.1.20);
> – is in compliance with applicable law and consistent with international norms of **behaviour** (2.1.10); and
> – is integrated throughout the **organization** (2.1.12) and practised in its relationships
>
> NOTE 1   Activities include products, services and processes.
> NOTE 2   Relationships refer to an organization's activities within its **sphere of influence** (2.1.19). (ISO 26 000, 2010)

It can be noted that the social responsibility definition focuses on the organization, which is the main subject of this social responsibility. In comparison, sustainable development is a matter for society. This is extensively explained in a specific article of the standard which aims to clarify the articulation between sustainable development and social responsibility. The ISO states that 'sustainable development is about meeting the needs of society', whereas social responsibility 'has the organization as its focus and concerns the responsibilities of an organization to society and the environment'. But the two concepts are closely linked since sustainable development sums up 'the broader expectations of society that need to be taken into account by organizations seeking to act responsibly'. This leads ISO 26000 to affirm that 'an overarching goal of an organization's social responsibility should be to contribute to sustainable development'. In addition, it argues that 'The decisions and activities of a socially responsible organization can make a meaningful contribution to sustainable development'. The standard also clarifies the fact that sustainable development does not amount to the sustainability of an organization, and that indeed the durability of some organizations might be incompatible with sustainable development: '[sustainable development] does not concern the sustainability or ongoing viability of any specific organization. The sustainability of an individual organization may, or may not, be compatible with the sustainability of society as a whole'. One can then understand how social responsibility should contribute to sustainable development, but that sustainable development cannot be reduced to the social responsibility of organizations.

The articulation between social responsibility and sustainable development in the ISO 26000 standard is:

3.3.5 Relationship between social responsibility and sustainable development
Although many people use the terms social responsibility and sustainable development interchangeably, and there is a close relationship between the two, they are different concepts.

Sustainable development is a widely accepted concept and guiding objective that gained international recognition following the publication in 1987 of the Report of the World Commission on Environment and Development: Our Common Future . . . Sustainable development is about meeting the needs of society while living within the planet's ecological limits and without jeopardizing the ability of future generations to meet their needs. Sustainable development has three dimensions – economic, social and environmental – which are interdependent; for instance, the elimination of poverty requires both protection of the environment and social justice.

Numerous international forums have reiterated the importance of these objectives over the years since 1987, such as the United Nations Conference on Environment and Development in 1992 and the World Summit on Sustainable Development in 2002.

Social responsibility has the organization as its focus and concerns the responsibilities of an organization to society and the environment. Social responsibility is closely linked to sustainable development. Because sustainable development is about the economic, social and environmental goals common to all people, it can be used as a way of summing up the broader expectations of society that need to be taken into account by organizations seeking to act responsibly. Therefore, an overarching goal of an organization's social responsibility should be to contribute to sustainable development.

The principles, practices and core subjects described in the following clauses of this International Standard form the basis for an organization's practical application of social responsibility and its contribution to sustainable development. The decisions and activities of a socially responsible organization can make a meaningful contribution to sustainable development.

The objective of sustainable development is to achieve sustainability for society as a whole and the planet. It does not concern the sustainability or ongoing viability of any specific organization. The sustainability of an individual organization may, or may not, be compatible with the sustainability of society as a whole, which is attained by addressing social, economic and environmental aspects in an integrated manner. Sustainable consumption, sustainable resource use and sustainable livelihoods relate to the sustainability of society as a whole. (ISO 26 000, 2010)

Based on the ISO 26 000 standard, social responsibility can be understood as the operationalization of sustainable development at the organization level. This articulation is particularly interesting because if sustainability has often been reduced to environmental issues in most organizations, social issues in ISO 26 000 are very well detailed and constitute a major part of the standard. In a way, the sustainable development concept has benefited from its recent association with social responsibility and the long history during which it has been discussed and analyzed.

Indeed the concept of social responsibility emerged way before environmental concerns arose, at the beginning of the century, to address the tension between corporate private goals and society's welfare and aspirations, and more specifically took an interest in the responsibility of corporations with regard to social issues such as labor, local communities, and so on. Social responsibility research has explored social expectations towards business, its responsibilities and its impacts. Anchored in different historical and ideological debates, the social responsibility literature has tended to explain why and how business should take social pressure and outcomes into account (Wood, 1991). Some scholars have insisted on the moral dimension of such responsibility, but most of them participated from an instrumental perspective, showing that social responsibility is fruitful to organizations, at least in the long run (Scherer and Palazzo, 2007: 1096). The 'business case' for social responsibility is that such positioning is a win–win (business and society), or a win–win–win (business, environment and society) strategy. The message has been that business should be socially responsible either to be more profitable (by gaining new markets, reducing risk, enhancing image or saving costs), to gain discretionary power (by avoiding new regulations), or at least to ensure its legitimacy (by increasing its social acceptance). Throughout the history of the social responsibility concept, what has been at stake is the autonomy

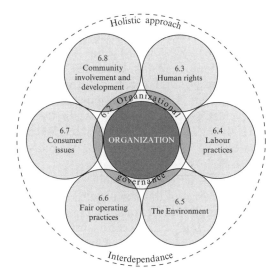

Source: ISO 26 000 (2010).

Figure 3.2   *An illustration of the seven core subjects of ISO 26 000 standard*

of business: does it need to be regulated to participate in social welfare, or can it self-regulate to maximize its contribution to society by voluntary social responsibility strategies? With the release of the ISO 26 000 standard, this debate has been overcome since it embodies the law as a constitutive element of social responsibility, instead of presenting the law and social responsibility initiatives as alternatives. But most important for this discussion is the fact that social responsibility research and practical organization experience provided the opportunity to develop knowledge about the social issues of business. This knowledge could fruitfully be articulated to the sustainable development concept, to detail what its social pillar can mean for organizations.

Therefore, the conceptual link proposed by ISO 26 000 is doubly interesting because the standard takes note that social expectations are now crystallized in the sustainable development concept. But it also proposes a richer understanding of the social issues raised by business compared to what can be tackled by the unidimensional social pillar of the sustainable development definition. The ISO 26 000 standard identifies seven core subjects (Figure 3.2), of which only one does not concern a social matter and refers directly to the environment. The others can all be understood as declinations of social or socio-economic issues: human rights, labor practices, consumer issues, community involvement and development, organizational governance, and fair operating practices.

The declared aim of the ISO standard is not to propose a management system to manage social responsibility, but rather to offer a clearer definition of what the social responsibilities of organizations are in the context of sustainable development and globalization. Nevertheless, it proposes some tools to implement social responsibility, firstly by a prescription about stakeholder dialogue and engagement (clause 5), and secondly by suggestions on how to integrate social responsibility throughout the organization (clause 7).

Given the changes that social responsibility strategies impose on the governance and management of the organization, one might be surprised at how little attention is given to innovation in the ISO 26000 standard. There are no more than four short mentions of innovation in different clauses. In the chapter devoted to fair operating practices, innovation is linked to fair and widespread competition as well as property rights: 'Fair and widespread competition stimulates innovation and efficiency . . . Recognition of property rights promotes investment and economic and physical security, as well as encouraging creativity and innovation'. Innovation is also presented as one of the possible benefits of social responsibility for an organization: 'Social responsibility can provide numerous benefits for an organization. These include: . . . generating innovation'. And finally, innovation is cited as an appropriate approach for environmental management activities: 'In its environmental management activities, an organization should assess the relevance of, and employ as appropriate, the following approaches and strategies: . . . *life cycle approach* . . . An organization should focus on innovations, not only on compliance, and should commit to continuous improvements in its environmental performance'.

These few references do not place innovation as the main concern of organizational modernization towards sustainable development in ISO 26000. In comparison, innovation is a fundamental piece of the green economy proposition. It structures the necessary transition of productive and consumption systems that sustainable development requires.

THE ENVIRONMENT–ECONOMY DEBATE: THE INNOVATION KEY

From the early stages of environmental consciousness, technology and innovation have been important pieces of the debate over environmental crisis and economic development. Notwithstanding the anti-technological current associated with specific factions of the ecologist movement (Yearly, 1994), technological innovation has been presented as a magic tool to reconcile economic growth and environmental conservation by dematerializing economic activities. It was invoked as the main response to the so-called alarmist discourses that opened the ecologist era, notably *The Population Bomb* by Ehrlich (1968) and the Meadows et al. report *Limits to Growth* (1972). The main argument is that those catastrophist scenarios are built on the hypothesis of a constant technology. But technology evolves and changes the impact that a population, even growing, has on the environment. The dynamic under this technological progress relies on an automatic economic adjustment, since the growing scarcity of a resource raises its price, and forces an efficiency strategy or the use of alternative resources guaranteeing its future conservation.[2] This dynamic stimulates innovation and ensures either the discovery of other resources, or more efficiency in the exploitation of the current resource.

But critics have pointed out that too much faith is placed in technology, which cannot totally dematerialize human needs, and which development responds to other logics than the utilitarian one to solve the ecological crisis. Moreover, the demand placed on technology seems unrealistic, if we follow Ekins' reasoning built on the Ehrlich and Ehrlich equation.

An equation introduced by Ehrlich and Ehrlich (1990: 58) indicates the scale of the technological challenge if both sustainability and gross national product (GNP) growth are to be achieved. They relate environmental impact (I) to the product of three variables: population (P), consumption per capita (C), and the environmental intensity of consumption (T). This last variable captures all the changes in technology, factor inputs, and the composition of GNP. Thus: $I = PCT$.

The contemporary concern with sustainable development indicates that current levels of I are unsustainable . . . the necessary reductions in T (T_R) in order to reduce environmental impacts to 50 percent of the current value by 2050 would be as follows . . .:

1. No growth in P or C in North and South $T_{R1} = 50\%$
2. Growth in P, no growth in C $T_{R2} = 65\%$
3. Growth in P and C in South $T_{R3} = 81\%$
4. Growth in P and C in North $T_{R4} = 89\%$
5. Growth in P and C in North and South $T_{R5} = 91\%$

These figures clearly illustrate some important aspects of the technology–sustainability relations. T_{R5} shows that, with moderate GNP growth in North and South, and projected population growth, the environmental impact of each unit of consumption would need to fall by 91 percent over the next 50 years to meet the rather conservative definition of sustainability that has been adopted. This is a very tall order indeed, and one does not have to be a technological pessimist to entertain serious doubts as to its feasibility. (Ekins, 1994: 130)

Moreover, technological innovation is not a determined process that can easily be planned, and it is anchored in a social dynamic that influences its path (Swaney, 1988: 344; Salomon, 1992: 45). Indeed, the interplay between actors and power relations influences technological development in such a way that it is impossible to ensure a best technology outcome (Godard and Salles, 1991: 254). It must also be noted that some technologies tend to transfer pollution instead of reducing or eliminating it (Swaney, 1988: 345), whereas others create entirely new and unpredicted problems (Duclos, 1993: 318–21), or result in a Jevon paradox or rebound effect, or a more general nemesis effect.

The Porter perspective on this issue has renewed the debate by adding the regulation factor to the economic dynamic. In a short essay published in 1991, Porter formulates what will be referred to as the Porter Hypothesis: stringent regulation fosters competitiveness by stimulating innovation. In response to what he sees as a narrow view of the sources of prosperity and a static understanding of competition, he builds on the fact that, as shown in his book *The Competitive Advantage of Nations*, 'the nations with the most rigorous requirements often lead in exports of affected products' (Porter, 1991: 168). He complains about the fact that United States (US) has lost its leadership in setting environmental standards. In comparison, Germany, which has some of the tightest regulation of air quality in the world, seems to lead in pollution control equipment on the world markets. Porter's main argument is that if standards can raise costs at the beginning, 'Properly constructed regulatory standards, which aim at outcomes and not methods, will encourage companies to re-engineer their technology. The result in many cases is a process that not only pollutes less but lowers costs or improves quality' (Porter, 1991).

This hypothesis is developed in a paper published four years later by Porter and van der Linde in the *Journal of Economic Perspectives*: 'Toward a new conception of the environment–competitiveness relationship' (1995). In the new competitiveness paradigm described in Porter's earlier work, innovation plays a central role: 'Competitive advantage . . . rests not on static efficiency nor on optimizing within fixed constraints, but on the capacity for innovation and improvement that shift the constraints' (Porter and van der Linde, 1994: 98). Instead of relying solely on an economic or market dynamic and responding to the worry that environmental protection measure will raise production costs, the authors evoke environmental regulation as a powerful innovation factor that enhances both environmental and competitive performance:

> properly designed environmental standards can trigger innovation that may partially or more than fully offset the costs of complying with them. Such

'innovation offsets', as we call them, can not only lower the net cost of meeting environmental regulations, but can even lead to absolute advantages over firms in foreign countries not subject to similar regulations. Innovation offsets will be common because reducing pollution is often coincident with improving the productivity with which resources are used. In short, firms can actually benefit from properly crafted environmental regulations that are more stringent (or are imposed earlier) than those faced by their competitors in other countries. By stimulating innovation, strict environmental regulations can actually enhance competitiveness. (Porter and van der Linde, 1994: 98)

In the authors' view, regulation is necessary because firms do not always make optimal choices, notably in an imperfect information context. But Porter and van der Linde do not mobilize market externalities to explain that resource efficiency or pollution reduction will not necessarily be rewarded by the market and in the cost structure of production. They do not present regulation as a mean to internalize environmental costs so that resource efficiency goes hand in hand with the reduction of production costs.

For Porter and van der Linde, regulation influences the direction of the innovation efforts of firms, and when properly crafted 'can serve at least six purposes' (Porter and van der Linde, 1994: 99). Firstly, regulation is a signal of 'resource inefficiency and potential technological improvements'. Secondly, when aimed at information gathering, it gives firms a clearer idea of their environmental performance. Thirdly, it guarantees that environmental investment will be valuable. Fourthly, in the same way as strong competitors can do, regulation creates pressure for innovation. Fifthly, it creates a level playing field for all organizations in order to avoid prisoner's dilemma behaviours. And finally, regulation is necessary when innovation offsets cannot compensate for environmental quality improvements.

The authors explain that when subject to stringent regulation, firms innovate in two ways. They gain knowledge and expertise in how to address environmental problems, or they entirely rethink their products and production process, which of course can lead not only to a reduction in compliance costs, but also to an enhancement of product quality and the production process.

Being in total contradiction with the conventional wisdom of that time, the Porter Hypothesis has provoked numerous reactions, still continuing today (Ambec et al., 2011: 1). As early as 1995 in their article, Porter and van de Linde took no less than four pages to respond to the critics provoked by Porter's 1991 essay. But as Ambec et al. note, after 20 years of debate since its publication:

on the theoretical side, it turns out that the theoretical arguments that could justify the PH [Porter Hypothesis] are now more solid than they appeared at first in the heated debate that took place in 1995 in the *Journal of Economic Perspectives* (Palmer et al., 1995). On the empirical side, on one hand, the evidence about the 'weak' version of the hypothesis (stricter regulation leads to more innovation) is also fairly well established. On the other hand, the empirical evidence on the strong version (stricter regulation enhances business performance) is mixed, with more recent studies providing more supportive results. (Ambec et al., 2011: 16)

In any case, those critics did not stop UNEP explicitly referring to the Porter Hypothesis in its Green Economy full report. At page 22, the report says:

> At the national level, any strategy to green economies should consider the impact of environmental policies within the broader context of policies to address innovation and economic performance (Porter and Van der Linde, 1995).[2] In this view, government policy plays a critical role within economies to encourage innovation and growth. Such intervention is important as a means for fostering innovation and for choosing the direction of change (Stoneman ed. 1995; Foray ed. 2009).
>   2. This point has been debated since at least the time of the initial statement of the Porter Hypothesis. Porter argued then that environmental regulation might have a positive impact on growth through the dynamic effects it engendered within an economy. (UNEP, 2011: 22)

What is striking in the *Green Economy* report, as in the Porter and van der Linde article, is the almost non-existence of the environmental externality argument as the grounds and justification for environmental regulation. Regulation is presented as a competitive tool, not as a necessary way to internalize environmental externalities. Absent from Porter's article, externalities are especially presented as a consequence of bad regulation and economic incentives in the UNEP report. They are also considered in the agriculture sector, but are never central to the justification of regulation for a green economy. This might seem surprising given that during recent decades, the core debate in environmental economics has focused on the existence and the necessary management of those externalities, from Pigou (1921) to Coase (1960), without missing the social cost concept of Kapp (1950). But this can be understood given the political context, where the economic vitality argument is of prime importance. The argument responds more directly to the economy–environment dichotomy perspective that seems to have slowed down environmental protection efforts in the context of a resurgent economic crisis.

This probably explains the importance that innovation has in the *Green Economy* report. Innovation occurs no less than ten times in the synthesis version, and is presented as a result of enabling conditions for a transition to a green economy (UNEP, 2011: 31). Innovation is cited in four out of the five reasons invoked to explain why the shift to a green economy 'might be good for long-term competitiveness as well as for social welfare' (UNEP, 2011: 22).

The UNEP report is directed at nations and policy-makers, but we can see that Porter Hypothesis has been transposed at the organizational level and seems finally to have become the new conventional management wisdom. An illustration of this is the special issue of the *Harvard Business Review* entitled 'Make Green Profitable' released in spring 2010. In addition to an article by Porter and Kramer on the link between competitive advantage and corporate social responsibility and many others, Nidumolu et al. propose a paper entitled 'Why sustainability is now the key driver of innovation' (Nidumolu et al., 2010: 78–86). Built on the idea that environment is an opportunity to lower costs and increase revenues instead of a costly constraint, they argue that: 'In the future, only companies that make sustainability a goal will achieve competitive advantage. That means rethinking business models as well as products, technologies, and processes' (2010: 81). They propose a five-stage process to integrate sustainability as an organization goal (2010: 82–3):

- Stage 1: Viewing compliance as opportunity;
- Stage 2: Making value chains sustainable;
- Stage 3: Designing sustainable products and services;
- Stage 4: Developing new business models;
- Stage 5: Creating next-practice platforms.

But in this article, as in the formulation of Porter's Hypothesis, the authors do not discuss the specificity of the innovation process in the context of, or as a tool for, sustainable development. Given the importance given to innovation in the ecological modernization process of the economy, there is a growing body of literature studying the specific challenge of innovation for sustainable development. In analyzing the conditions under which technology could contribute to solve the environmental problem, those studies propose what can be viewed as a concrete answer to the critics of the fact that too much faith is placed in technology.

INNOVATION FOR SUSTAINABLE DEVELOPMENT

In an interesting article published in 2002, Vollenbroek (2002) questions the challenge of innovation in the sustainable development context. He argues that innovation process must be oriented in new ways to ensure that it contributes to the quality of life and well-being of populations. We can no longer rely on the dynamic between economic agents who concur on an innovation path building on their understanding of the larger context and its opportunities. Given the challenges we face as a society, we need to direct the innovation process towards our needs in an approach which Vollenbroek qualifies as 'society pull'. This approach is referred to as transition management: 'a process approach directing innovation towards sustainable development' (Vollenbroek, 2002: 215). Its main characteristic is that 'innovation is no longer driven by the past', but directed towards a shared vision of the future (Vollenbroek, 2002).

This new orientation is not easy to achieve, and requires a cultural change towards the innovation process. Since the Enlightenment, science and technology have been considered as powerful means to improve the quality of life. Given this positive role in society, governments have proposed subsidies to stimulate innovation, letting the market dynamic choose which technologies would be kept, and which abandoned. But as explained above, sustainable development imposes new expectations on technology. And if we are to orient technological development to contribute directly to sustainable development goals, generic innovation subsidies will not be sufficient, as shown by earlier experiences:

> the programme EET (Economy, Ecology, Technology) was developed in which economical and ecological objectives meet in one subsidy scheme. Although this scheme is very successful, it is not expected to be sufficient to reach the ultimate goal: substitution of current technologies by new environmentally benign technologies or product-service combinations. In order to reach this goal, policy coherence between economic, science, technology and industry policy is needed, thus contributing to the formation of an innovation system that is directed towards sustainable development and which creates opportunities for producers and consumers. (Vollenbroek, 2002: 216)

The coherence that Vollenbroek is proposing leads indeed to the idea that 'system changes are required, which will enable the fulfillment of needs in an entirely new manner' (2002: 217). But it also supposes a new development dynamic of technology, called 'society pull', as illustrated by the Dutch program for Sustainable Technology Development experience:

> The programme also demonstrated that such achievements [the required improvement in eco-efficiency] are possible by using future visions to derive the R&D [research and development] agenda of today. Another important element of the programme is the co-evolution of technology, structure and culture. This means that not only technology should be considered as a driver for innovation, but also societal needs and goals must also [*sic*] be factored in. Obviously this means that public policy-makers should be able to create conditions which are

attractive enough for private parties to co-operate. On the other hand, private parties should be prepared to commit to public goal. (Vollenbroek, 2002: 217–18)

Government has a central role in this society pull dynamic, called transition management, that impels a 'process approach directing innovation towards sustainable development' (2002: 215). To manage this transition requires, firstly, a deep understanding of the actors involved; allowing, secondly, that the transition objective be formulated in a shared perspective. And to ensure private parties' participation in the process, explains Vollenbroek, it is necessary that they see business opportunities in it (2002: 219).

As we can see, innovation for sustainable development needs to be thought of more than simply as a natural self-driven solution to the environment–economy dichotomy, be it stimulated by market dynamics or by stringent regulation. It is indeed a social process that needs to be managed in a more conscious way if we are to rely on it to build a green economy and more generally attain social goals. But it is not only technological development that we are talking about: we need to rethink the way our needs are defined, and understand more closely the fact that they are socially constructed. Vollenbroek explains how technological development must be thought of in relation to needs, that are social constructions, and argues that 'the present generation should develop knowledge, which can be used by next generations to construct and meet their needs, rather than focusing on saving some non-renewable resources' (2002: 217). This knowledge might help next generations to fulfill their needs while depending less on the environment, or if we prefer, to fulfill their needs in a dematerialized way, giving an important role to culture, shared beliefs and the societal paradigm. This perspective is interesting in thinking about the innovation process, because it sheds light on social innovation, rather than on technologies to which innovation has traditionally been associated. But it also gives a predominant role to civil society actors who have a direct impact on the culture of a society, the way challenges are understood, and how desires are shaped. In this respect, non-governmental organizations (NGOs) can contribute directly to those transformations by sharing their knowledge, assessing technologies and forming coalitions with selected private actors.

This is why some scholars propose the concept of a socio-technical regime to capture the transformations at stake when talking about innovation for sustainable development:

The term 'socio-technical regime' captures this complex configuration of artefacts, institutions, and agents reproducing technological practices. The socio-technical 'adjective is used to stress the pervasive technological mediation of social relations, the inherently social nature of all technological entities, and indeed the arbitrary and misleading nature of distinctions between "social" and "technical": elements, institutions or spheres of activity' (Russel & Williams, 2002: 128). The development of the socio-technical is a highly social, collective process, and ultimately it is diverse social actors who negotiate innovation (Smith et al., 2005). Imposing a normative goal like sustainable development upon existing socio-technical regimes implies connecting and synchronizing changes amongst actors, institutions and artefacts at many different points within and beyond the regime. (Seyfang and Smith, 2007: 588)

In their article, Seyfang and Smith explain that given the challenge of sustainable development, the niche innovation model must be completed at least by what they call grass-roots innovations which respond to a totally different logic of development. While niches are special spaces within the market where the rules are different, grass-roots innovations develop in 'the social economy of community activities and social enterprise' (2007: 591). But, as they show, 'the grassroots is a neglected site of innovation for sustainable development' even if it has strong potential (2007: 598). The innovation process for sustainable development certainly requires that more attention is paid to grass-roots experiments. Moreover, organizations might gain from a direct dialog with civil society, as recent research shows. Ayuso et al. (2006) studied how firms can integrate 'stakeholder insights into the process of organizational innovation from a sustainable development viewpoint' by studying two cases where stakeholder dialogue helped generate innovations that were beneficial to the firm as well as for sustainable development:

The evidence from the two case studies suggests the existence of two simple capabilities – stakeholder dialogue and stakeholder knowledge integration – for generating innovations in accordance with stakeholders needs. Whereas stakeholder dialogue leverages organizational resources that promote two-way communication, transparency and appropriate feedback to stakeholders, stakeholder knowledge integration relies on non-hierarchical structures, flexibility and openness to change. The paper sheds some light on the under-researched issue of linking stakeholder dialogue and sustainable innovation, and thus contributes to opening the 'black box' of dynamic capabilities and advancing in the understanding of this fundamental organizational concept. (2006: 1)

Even if we can see interesting developments in the ecological modernization of production processes, it seems that the dialogue between social and technical innovations is not even planned in many cases. For example, we could start a dialog between eco-design (design for the environment) and co-creation (mutual design by the producer and the user) processes, that are still very independent one from another in their practice as well as in their theorization. And there are numerous other cases where social actors would be tremendously helpful in redesigning not only products, but the very definition of the needs they pretend to fulfill. On the other side, information and education can participate in the reshaping of these needs and promote a more responsible and ecological consumption.

CONCLUSION

Sustainable development will require multiple adjustments and reorganizations, not only in our public policies and in business strategies, but even in the way we understand its challenge and try to respond to it. Relying on innovation to respond to sustainable development challenges means that we pretend to manage this very specific process of innovation, which was thought of only few years ago as an autonomous and undetermined phenomenon. If we are to succeed in this transition management, not only the state and firms will be central, but also NGOs playing an active role as they possess a different perspective and knowledge that can be valuable specifically from an innovation point of view (Seyfang and Smith, 2006: 592–5).

But if the cultural transition allowing us to better manage our relation to a limited and fragile biosphere seems possible, the time to accomplish such social change might be insufficient given how long it takes for social dynamics, structures and beliefs to evolve. In this context, the consciousness of the environmental crisis is determinant, but so is the understanding of the evolution and transformation of social, economic and political systems.

NOTES

1. The ISO 26 000 citation comes from the DIS version ISO/TMB/WG SR N 172 DRAFT INTERNATIONAL STANDARD ISO/DIS 26 000.
2. But it has been demonstrated that this dynamic does not lead to an ecological optimum because of the differing moments of ecological and economic adjustments.

REFERENCES

Ambec, S., M.A. Cohen, E. Stewart and P. Lanoie (2011), 'The Porter Hypothesis at 20: can environmental regulation enhance innovation and competitiveness?', Discussion Papers dp-11-01, Resources For the Future.
Ayuso, S., M.A. Rodrıguez and J.E. Ricart (2006), 'Responsible competitiveness

at the "micro" level of the firm: using stakeholder dialogue as a source for new ideas: a dynamic capability underlying sustainable innovation', *Corporate Governance*, **6**(4): 475–90.

Coase, R.H. (1960), 'The problem of social cost', *Journal of Law and Economics*, **3**: 1–44.

Daly, H.E. (1990), 'Toward some operational principles of sustainable development', *Ecological Economics*, **2**: 1–6.

Duclos, D. (1993), 'La dérive technologiste', in M. Beaud, C. Beaud and M.L. Bouguerra (eds), *L'état de l'environnement dans le monde*, Paris: La Découverte, pp. 318–22.

Ehrlich, P. (1968), *The Population Bomb*, New York: Ballantine Books.

Ehrlich, P.R. and A.H. Ehrlich (1990), *The Population Explosion*, New York: Simon & Schuster.

Ekins, P. (1994), 'Sustainable development and the economic growth debate', in B. Bürgenmeier (ed.), *Economy, Environment, and Technology: A Socio-Economic Approach*, New York: M.E. Sharpe, pp. 121–37.

Faucheux, S. (1995), 'Quels indicateurs choisir pour évaluer la durabilité?' *Écodécision*, **15**: 64–5.

Gendron, C. and J.P. Revéret (2000), 'Le développement durable', *Économies et Sociétés*, **F**(37): 117–24.

Godard, O. and J.M. Salles (1991), 'Entre nature et société: les jeux de l'irréversibilité dans la construction économique et sociale du champ de l'environnement', in R. Boyer, B. Chavance and O. Godard (eds), *Les figures de l'irréversibilité en économie*, Paris: Éditions de l'École des Hautes études en sciences sociales, pp. 233–72.

Goodland, R. (1995), 'The concept of sustainability', *Écodécision*, **15**: 30–32.

ISO 26000 (2010), *Guidance on Social Responsibility*.

Kapp, W.K. (1950), *Social Costs of Private Enterprise*, Cambridge, MA: Harvard University Press.

Meadows, D.H., D.L. Meadows, J. Randers and W.W. Behrens III (1972), *The Limits to Growth*, New York: Universe Books.

Nidumolu, R., C.K. Prahalad and M.R. Rangaswami (2010), 'Why sustainability is now the key driver of innovation', *Harvard Business Review*.

Palmer, K., W.E. Oates and P.R. Portney (1995), 'Tightening environmental standards: the benefit–cost or the no-cost paradigm?' *Journal of Economic Perspectives*, **9**(4): 119–32.

Pearce, D.W. and G.D. Atkinson (1993), 'Capital theory and the measurement of sustainable development: an indicator of "weak" sustainability', *Ecological Economics*, **8**: 103–8.

Pigou, A.C. (1921), *The Economics of Welfare*, London: Macmillan.

Porter, M. (1991), 'America's green strategy', *Scientific American*, **264**(4): 168.

Porter, M. and C. van der Linde (1995), 'Toward a new conception of the environment–competitiveness relationship', *Journal of Economic Perspective*, **9**(4): 97–118.

Rist, G. (1996), *Le développement: histoire d'une croyance occidentale*, Paris: Presses de la Fondation nationale des sciences politiques.

Rostow, W.W. (1960), *Les étapes de la croissance économique: un manifeste non communiste*, Paris: Seuil.

Russell, S. and R. Williams (2002), 'Social shaping of technology: frameworks, findings and implications for policy with glossary of social shaping concepts', in K.H. Sørensen and R. Williams (eds), *Shaping Technology, Guiding Policy: Concepts, Spaces and Tools*, Cheltenham, UK and Northampton, MA, USA: Edward Elgar, pp. 37–132.

Salomon, J.J. (1992), *Le destin technologique*, Paris: Balland/Gallimard.

Scherer, A.G. and G. Palazzo (2007), 'Toward a political conception of corporate responsibility: business and society seen from a Habermasian perspective', *Academy of Management Review*, **32**(4): 1096–120.

Seyfang, Gill and Adrian Smith (2007), 'Grassroots innovations for sustainable development: towards a new research and policy agenda', *Environmental Politics*, **16**:4, 584–603.

Seyfang, G. and A. Smith (2007), 'Grassroots innovations for sustainable development: towards a new research and policy agenda', *Environmental Politics*, **16**(4): 584–603.

Smith, A., A. Stirling and F. Berkhout (2005) 'The governance of sustainable socio-technical transitions', *Research Policy*, **34**: 1491–510.

Swaney, J.A. (1988), 'Elements of a neoinstitutional environmental economics', in M.R. Tool (ed.), *Evolutionary Economics: Institutional Theory and Policy*, Vol. 2, New York: M.E. Sharpe, pp. 321–61.

UN (1972), *Stockholm Declaration on the Human Environment*.

UNEP (2011), *Towards a Green Economy: Pathways to Sustainable Development and Poverty Eradication – A Synthesis for Policy Makers*.

Vollenbroek, F.A. (2002), 'Sustainable development and the challenge of innovation', *Journal of Clean Production*, **10**: 215–23.

Wood, D.J. (1991), 'Corporate social performance revisited', *Academy of Management Review*, **16**(4): 691–718.

Yearley, S. (1994), 'Social movements and environmental change', in Michael Redclift and Ted Benton (eds), *Social Theory and the Global Environment*, London, UK and New York, USA: Routledge: pp. 150–168.

Web References

HSBC (2009), 'A climate for recovery – the colour of stimulus goes green', HSBC Climate Change Global Research, available at http://www.globaldashboard.org/wp-content/ uploads/2009/HSBC_Green_New_Deal.pdf (accessed 5/06/2012).

4. Appraisal of corporate governance norms: evidence from Indian corporate enterprises

Rabi Narayan Kar

INTRODUCTION

The issues relating to corporate governance have come into prominence because of their apparent importance for the sustainable health of corporations and society in general, especially after the plethora of corporate scams and debacles in recent times. The United States (US), Canada, United Kingdom (UK), other European countries, East Asian countries, and even India for that matter have witnessed the collapse of or severe pressure on their economies and have faced grave problems including the demise of several leading companies in the recent past. This has resulted in a greater emphasis on and new dimensions of corporate governance issues. The corporate governance issues flow from the concept of accountability for the safety and performance of assets and resources entrusted to the operating team. These issues of accountability and governance assume greater significance and call for more sustainable practices in the case of developing countries.

Corporate governance broadly refers to a set of strategies and practices that are designed to govern the behaviour of corporate enterprises. Corporate governance deals with the ethos, laws, procedures, practices and implicit rules that determine a company's ability to take managerial decisions and innovative strategies towards sustainable growth. Against the backdrop of several corporate debacles, corporate governance has been increasingly seen as a means to promote healthier and sustainable corporate practices. In this context, governments all over the world are playing the role of facilitators by promoting good corporate governance norms for the ethical and sustainable functioning of enterprises. This chapter attempts to explore various dimensions of the regulatory framework of corporate governance in India. The case study of Satyam Computer Services Ltd has been incorporated to analyse the issues and draw comparisons with other corporate failures, and their implications.

EVOLUTION OF INDIAN ENTERPRISES AND CORPORATE GOVERNANCE ISSUES

Pre-Independence Period

In 1809, the first managing company was established in India. Carr Tagore & Company (1834) was known as the first equal partnership between Indian and European businessmen, initiating the managing agency system in India. In 1850, the Act for the registration of the companies with limited liabilities was passed. The Companies Act was introduced in the year 1866 and was further revised in 1882, 1913 and 1932. The Indian Partnership Act was introduced for the first time in 1932. The main emphasis during that period was on the managing agency model of corporate affairs as individuals and business firms entered into legal contracts with joint stock companies. The model was characterized by the abuse and misuse of responsibilities by managing agents due to dispersed ownership. The issues of profit generation and control were dilapidated, leading to various conflicts.

Further, mergers and acquisitions have played an important role in the transformation of the Indian corporate sector since the Second World War period. The economic and political conditions during the Second World War and post-war periods (including several years after independence) gave rise to the expansion of Indian companies through a spate of mergers and acquisitions (Kothari, 1967). The inflationary situation during wartime enabled many Indian businessmen to amass income by way of high profits and dividends and money made on the black market.

This led to wholesale entry of small businessmen and traders into industry during the war period giving rise to hectic activity on the Indian stock exchanges. There was a push to acquire control over industrial units in spite of inflated share prices. The practice of cornering shares in the open market and the trafficking of managing agency rights with a view to acquiring control over the management of established and reputed companies came prominently to light. The net effect of these two practices, that is, of acquiring control over ownership of companies and over managing agencies, was that a large number of concerns passed into the hands of prominent industrial houses of the country. As it became clear that India would be gaining independence, British managing agency houses gradually liquidated their holdings at fabulous prices offered by the Indian business community. Besides the transfer of managing agencies, there were a large number of cases of transfer of interests in individual industrial units from British to Indian hands. Further, at that time it was the fashion to obtain control of insurance companies, for the purpose of utilising their funds to acquire substantial holdings in other companies. The big industrialists also floated banks and investment companies for furtherance of the objective of acquiring control over established concerns.

Post-Independence Period

Independent India adopted its Constitution in 1950. The Directive Principles of State Policy of the Indian Constitution mandate that the State should work to prevent the concentration of wealth and means of production in a few hands, and try to ensure that ownership and control of material resources is distributed to best serve the common good (Article 38, Constitutional Law of India). The Gandhian trusteeship philosophy is based on an understanding that everything is owned by everyone, and wealth is owned by those who generate it. Thus, the person who controls an asset is not an owner but a trustee, being given control of that asset by the society.[1] These aspects show the genuine concern for adopting ethical and sustainable practices in Indian business governance. However, the following developments added a new dimension to the issues of corporate governance in the post-independence period.

Large numbers of companies were created through mergers and acquisitions in industries such as jute, cotton textiles, sugar, insurance, banking, electricity and tea plantations. However, the anti-big-government policies and regulations of the 1960s and 1970s seriously deterred the expansion of Indian companies. The deterrent was mostly to horizontal combinations which result in concentration of economic power to the common detriment. This does not, of course, mean that mergers and acquisitions were uncommon during the controlled regime. There were many conglomerate combinations. In some cases even the government encouraged mergers and acquisitions; especially for ailing units. Further, the formation of the Life Insurance Corporation and nationalization of the life insurance business in 1956 resulted in the takeover of 243 insurance companies. There was a similar development in the general insurance business. The National Textiles Corporation (NTC) took over a large number of ailing textiles units.

Further, between 1951 and 1974, a series of governmental regulations was introduced for controlling the operations of large industrial organizations in the private sector. Such regulations influenced considerably the growth strategies adopted by the companies. Some of the important regulations were: the Industries Development and Regulation Act, 1951; Indian Companies Act, 1956; Import Control Order, 1957–58; Monopolies and Restrictive Trade Practices Act, 1969; and Foreign Exchange Regulation

Act, 1973. The government abolished the managing agency system in 1970 and this period is also remembered for the nationalization of Indian banks. These regulations, along with others, influenced the governance and functioning of the Indian corporate enterprises.

Post-1990

Till 1980, the majority of big corporate houses in India were run by businesses, families and were established by first-generation entrepreneurs. The biggest and most successful business houses such as Tata, Birla and Bajaj voluntarily practised all good corporate governance norms. In spite of the absence of any express code for corporate governance, there was no major failure on governance issues in these business houses. The corporate management scenario started changing in India after the introduction of a liberalization process in 1991. Government regulations on companies relating to growth and expansion were reduced. Several measures taken by the government – which include delicensing, dereservation, enactment of the Securities Exchange Board of India Act (SEBI Act, 1992), Monopolies and Restrictive Trade Practices Act (MRTP Act, 1969) relaxations, and amendments to various financial and industrial regulations paving the way for liberalization of policy towards foreign capital and technology – led to a structural transformation in the Indian corporate sector (Das, 2000). This transformation has provided a launch pad for the corporate enterprises to grow and expand (Roy, 1999). After liberalization, India has been looked at by the developed markets for the purpose of creating new markets. This has led to the emergence of competitive forces in the Indian market (Basant, 2000). In conformity with global practices, progressive companies in India have made an attempt to put systems of good corporate governance in place in order to strike a balance between profit and sustainable growth. These developments have driven the corporate enterprises and the regulators towards the development of a system of corporate governance in India.

THE INDIAN INITIATIVES IN CORPORATE GOVERNANCE

The Indian Companies Act 1956 which replaced the Indian Companies Act 1913 has the regulatory mandate to control the affairs of companies. The Confederation of Indian Industries (CII) in 1996 took the major initiative of studying the corporate governance practices in India. The stated objective was to develop and promote a code for corporate governance.

The initiative by the CII flowed from public concerns regarding the protection of investor interests, especially those of the small investors; the promotion of transparency within business and industry, and the need to move towards international standards in terms of disclosure practices for sustainable growth of the corporate sector. Further, the focus was to develop a high level of societal confidence in business and industry. This resulted in the CII Code of corporate governance (CII, 1999).

The CII committee was followed by a committee set up by the Securities and Exchange Board of India (SEBI) under the chairmanship of Kumar Mangalam Birla (Birla Committee). The crux of the Birla Committee's report (SEBI, 1999) is the set of recommendations which distinguishes the responsibilities and obligations of the board and the management in instituting the systems for good corporate governance, and emphasizes the rights of shareholders in demanding good governance practices. Many of the recommendations are mandatory. The companies will also be required to disclose separately in their annual reports, a report on corporate governance delineating the steps they have taken to comply with the recommendations of the committee.

Further, the Securities and Exchange Board of India (SEBI), in its effort to improve the governance standards, constituted a committee under the chairmanship of N.R. Narayan Murthy (chairman, Infosys Technologies) to study the role of independent directors, related parties, risk management, directorship and director compensation, codes of conduct and financial disclosures in 2003. The recommendations of the committee have been incorporated in the clause 49 of the listing agreement.

In 2002, the Department of Company Affairs constituted a committee under the chairmanship of Naresh Chandra (the Naresh Chandra Committee) to, inter alia, examine the entire gamut of issues pertaining to the auditor–company relationship with a view to ensuring the professional nature of the relationship. In this context, the focus was on considering issues relating to the rotation of auditors and auditing partners, restrictions on non-audit fees, and procedures for the appointment of auditors, to ensure that the management and auditor present a true and fair view of the state of affairs of the company. It is also reflected in other measures such as the certification of accounts, and financial statements by the management and directors. The committee intended to study and build upon its report following the benchmarks set by the Sarbanes–Oxley Act. The recommendations of the committee have been adopted by the government.

In 2002, SEBI also set up a committee under the chairmanship of Mr Y.P. Malegam (the Malegam Committee) to review the prevailing disclosure requirements in the offer documents and suggest modifications with a view to making disclosures more transparent.

The Reserve Bank of India (RBI) had set up a consultative group of directors of banks and financial institutions under the chairmanship of Dr A.S. Ganguly (the Ganguly Committee, 2002) to review the supervisory role of boards of banks and financial institutions and to obtain feedback on the functioning of the boards vis-à-vis compliance, transparency, disclosures, audit committees, and so on. The Ganguly Committee report discussed several issues relating to the corporate governance of banks and financial institutions. The most important aspect of the report is the recommendation of a deed of covenant to be entered into by a person becoming a director of the company.

Furthermore, in 2004, a committee under the chairmanship of Dr J.J. Irani (the Dr Irani Committee) was set up to advise the government on the concept paper on company law, which included identifying the essential ingredients to be addressed by the new law, retaining desirable features of the existing framework, and segregating substantive law from the procedures to enable a clear framework for good corporate governance that addresses the concerns of all stakeholders equitably. It was expected to advise on enabling measures to protect the interests of stakeholders and investors, including small investors, for sound corporate governance practices. This committee also gave its recommendation on some of the aspects of corporate governance mechanisms such as the number of directors, the independence of directors, and so on.[2]

REGULATORY FRAMEWORK FOR CORPORATE GOVERNANCE

Article 38 of the Constitutional Law of India directs government to ensure equitable distribution of wealth. The clause states that government should work to prevent the concentration of wealth and means of production in a few hands, and try to ensure that ownership and control of the material resources is distributed to best serve the common good, which reflects the desire of the state for the adoption of good corporate governance practices. Companies in India are governed by the Companies Act, 1956 as amended. The Companies Act is administered by the Ministry of Corporate Affairs (MCA) and enforced by the Company Law Board (CLB) and the Company Courts.[3] Listed companies must comply with the rules and regulations prescribed by the Securities and Exchange Board of India (SEBI) Act, 1992; with the Securities Contract (Regulation) Act, 1956; the Depositories Act, 1996; the Sick Industrial Companies (Special Provisions) Act (SICA), 1985; and the listing rules. SEBI regulates the stock exchanges, stockbrokers, share transfer agents, merchant banks, portfolio managers, other market intermediaries, collective investment schemes and primary issues. It prohibits fraudulent and unfair trade practices, and regulates the substantial acquisition of shares and takcovers.[4] Over the last few years, a series of joint corporate governance committees were appointed by MCA and SEBI. Their recommendations are reflected in Companies Act amendments, listing rules and SEBI regulations (Sampath, 2006). Clause 49 of the listing rules addresses corporate governance on a 'comply or explain' basis.[5] Two credit rating agencies rate the quality of corporate governance of issuers.[6]

Companies have to follow the mandates of relevant Accounting Standards in preparing and reporting financial statements which are the best indicators to report how corporate governance principles are executed. These statements are prepared on the basis of Generally Accepted Accounting Principles (GAAP). Accounting Standards prescribe the recognition, valuation, reporting and disclosure of financial information.[7] Globalization has opened new horizons for businesses to expand their operations but at the same time global transactions raise new challenges. The parties to a business transaction often take undue advantage from international business. Transactions amongst associated concerns and relatives require special scrutiny. The Indian Income Tax Act, 1961 has enacted provisions to assess the true value of transactions by incorporating principles of transfer pricing.[8] By and large, express provision for corporate governance exists for listed companies only. In 2007, central government issued guidelines on corporate governance for central public enterprises. These are voluntary in nature and there are no similar guidelines for state-controlled public sector units as such. Similarly, clause 49 applies to the listed companies and there are no express provisions on corporate governance issues for unlisted companies irrespective of the size of the company. Accounting Standards issued by the Institute of Chartered Accountants of India (ICAI) have control over its members if they fail to comply in audit reporting. But the ICAI has no authority to act against enterprises violating Accounting Standards.[9] However, the Accounting Standards prepared and issued by the ICAI are mandatory only for its members, who, in discharging their audit function, are required to examine whether the said standards of accounting have been complied with.[10] Although India has a systematic regulatory framework for corporate governance encompassing various corporate legislations, yet the existing regulatory framework is being put to the test due to globalization, the competitive environment, increasing social concerns, and financial and economic reforms all over the world.

CORPORATE GOVERNANCE MECHANISM: A CASE STUDY OF SATYAM COMPUTER SERVICES LTD

The startling confessional disclosure by the promoter-chairman of Satyam Computer Services (India's fourth-largest and one of the most reputable information technology companies) that the company's accounts had been falsified for several years by the inflation of revenues and profit figures, exaggeration of cash and bank balances, overstatement of receivables and understatement of dues payable by the company, resulting in fraud of more than $1 billion, has sent shock waves all around the corporate world. The MCA has swung into action and sought permission from the CLB to suspend the Satyam board. The MCA has reportedly instructed the Registrar of Companies to examine the company's books and documents, and other sister concerns, to unearth the magnitude of the fraud, falsification of accounts and the route of the diverted resources. The MCA has also instructed the Serious Fraud Investigation Office (SFIO) to carry out investigations into the fraudulent dealings by Satyam Computer Services and to submit the report within three months.

Further, SEBI has started its own investigations to unearth the scam. Simultaneously, SEBI has reportedly decided to carry out a peer group audit of some of the important listed companies with a mandate to focus on inter-group company dealings. The ICAI has also sought an explanation from PricewaterhouseCoopers (PWC), the audit firm which has carried out the audit of Satyam Computer Services, and required it to furnish audit process details and documents so that responsibility for the lapses can be ascertained.

Satyam Computer Services Ltd was established in 1987. The company operates in the area of information technology (IT) services, and is listed on the New York Stock Exchange and Euronext as well as Indian exchanges. Satyam's network covers 67 countries across six continents and it employs 53 000 IT professionals across software development centres in India, the United States, the United Kingdom, Australia, the United Arab Emirates, Canada, Hungary, Singapore, Malaysia, China, Japan and Egypt. It provides services to over 654 global companies, 185 of which are Fortune 500 corporations. Satyam has several subsidiaries and group companies including the controversial Maytas Infrastructure. Satyam had failed in its attempt to buy Maytas in December 2008 which is engaged in the altogether different business line of real estate and infrastructure development. Maytas is controlled by Ramalingam Raju and his family members. Investigators have found that Mr Raju and his family members have floated 327 front companies, mainly having dealings in real estate.

In January 2009, Mr Raju confessed in a letter that the actual cash ($64.2million) was only 5 per cent of what was shown ($1063 million) on the balance sheet. It is shocking that top global auditing firm PricewaterhouseCoopers signed off Satyam's financial statements for a couple of years and that regulators in India, Europe and the United States apparently failed to take notice of problems in the company. However, Mr Raju denied that any other member of the board or family is involved or has knowledge of the scam. It is in fact unbelievable that a fraud of such magnitude can be carried for so long without the knowledge of key officials. Further, there has been widespread criticism directed at the independent directors of Satyam Computer Services. Either the independent directors have failed in their role or are inefficient.

The case narratives attempt to achieve the following pedagogical objectives: (1) to critically examine the role and responsibility of independent directors in the functioning of the company; (2) to investigate the role of the statutory auditors.

The Beginning of the Fall

The meaning of *satyam* in various Indian languages is 'truth'. However, the facts which have been emerging since December 2008 show that the company's financial statements reflected the opposite of the word *satyam*. The scam was unearthed when the chairman of the company, Mr B. Ramalingam Raju, led a move in the Satyam board meeting to use $1552 million of Satyam money to buy Maytas. The official justification for the move was strategic diversification. However, Indian public opinion including that of experts was unanimous in questioning the real motive for the acquisition of Maytas. The board of Satyam that took the controversial decision includes independent directors of high public standing. The widespread public anger forced the company to call off the move. Satyam had a majority of independent directors. Five of the nine directors are listed as being 'independent of management'. Satyam may have divided the roles of chairman and chief executive officer (CEO), but the two individuals who served in these roles were brothers and both were members of the top management. When unethical executives become intent on defrauding a company, it can be tough for even the most seasoned directors to see through the scam. But what is shocking in this case is the recognition Satyam has received for its governance standards, even though its board failed to adopt some basic and widespread boardroom best practices, some of which might well have triggered board awareness that something was amiss.

Transgression of Corporate Governance Norms

Following his confession on 7 January 2009, the prosecuting agencies arrested Mr Raju on 9 January. During interrogation, the chairman and chief finance officer (CFO) have reportedly confessed to the charges of falsifying the accounts for a number of years and siphoning off company money through salaries of more than 13 000 bogus employees. The Raju brothers have been booked for criminal breach of trust, cheating, criminal conspiracy and forgery under the Indian Penal Code (IPC), which are non-bailable offences and could put them behind bars for years if proved.[11]

Apart from the charges levelled by Indian government, investors in the USA have also filed a class lawsuit. Before the scam was revealed, the World Bank banned the errant firm from business for seven years.[12] The recent crisis surfaced when Satyam Computer Services Ltd proposed to acquire a stake in a company related to its promoters but unrelated to its core competency. This has called into question the role of independent directors on the board. Shareholders who rely on the presence of such directors to provide protection against transgressions of governance must now be wondering whether there may be other instances which have passed unnoticed.

Board composition

In the corporate form of business organization, the board of directors occupies a unique position. To have a board of directors is a legal requirement for all incorporated entities, mandated by statute. The Cadbury Report (1992) placed the corporate board at centre stage of the governance system, and described it as the means by which companies are directed and governed. Elected by the equity shareholders of the company, the board presides over the functioning and performance of the company, operates through the executive management, and is accountable to the shareholders and, in a broader sense, to the other stakeholders of the company. The corporate governance literature in the US and the UK focuses on the role of the board as a bridge between the owners and the management (Cadbury Committee, 1992). In an environment where ownership and management are widely separated, the owners are unable to exercise effective control over the management or the board. The management becomes self-perpetuating and the likes and dislikes of the CEO largely influence the composition of the board itself. The corporate governance reforms in the US and the UK have focused on making the board independent of the CEO. In India, guidelines on the composition of the board of directors have been issued along similar lines as abroad, mandating the appointment of a certain percentage of independent directors.[13] The role of independent directors poses a series of questions concerning their independence and the relationship of the board composition and independence with the firm's performance.

It has been observed that the composition of the board itself is largely influenced by the likes and dislikes of the chief executive officer, and hence the spirit of appointing independent directors as part of good corporate governance in practice serves no basic purpose in providing guidance in the overall interest of the company. Of the six independent directors serving on the Satyam board, four were academics, one was a former Cabinet Secretary of the Indian government, and only one (Vinod K. Dham) was a former chairman and CEO of a tech company (he had previously served as vice-president of Intel's Micro Processing Products Group).[14] Two of the independent directors, Mr Vinod K. Dham and Mr T.R. Prasad, are each noted in the 2008 SEC filing as serving on eight boards in addition to Satyam's.[15]

Board independence
Given the fiduciary relationships that the corporate directors are subject to, there is always an overwhelming need to ensure that they discharge their responsibilities properly to protect and promote the interests of shareholders as well as other stakeholders. It is in this context that measures to have independent directors on the board, who have no pecuniary relationships that may impair their judgment on matters relating to the company and its shareholders, are being stressed. The Cadbury Committee (1992), the Greenbury Committee (1995), the Hampel Committee (1998) and the Higgs Committee (2003) have mandated independent directors on the board. Then, the basic issue is to what extent independent directors are really independent in the decision-making process. There is no doubt that the appointment of independent directors on the boards of listed companies is to ensure adherence to good a corporate governance mechanism. Further, the independent directors have the ability to exercise checks and balances over attempted malpractices by the promoter-directors in illegally diverting funds of the company to entities belonging to the promoter family groups, and if the independent directors carry out their assigned task with great care and caution then they can question and scrutinize proposed deals and go behind the façade. Though they may not be able to unearth fraud and wrongdoings in the accounting records, which a professionally appointed auditor may do, yet the very presence of eminent outside experts of great stature may instil confidence in the investors and the regulatory authorities about the efficacy of good corporate governance.

Under the Indian Companies Act, 1956 there is no distinction between full-time executive directors and independent directors and all of them come under the definition of 'officer', as per section 2(30) of the Companies Act, 1956. Further, under section 291 of the Companies Act, the board of directors of the company is empowered to exercise all powers in the management of the affairs of the company and no limitation has been put on the exercise of powers by the company directors, whether they are independent directors or executive directors. Though the Companies Act, 1956 enables the appointment of a managing director who exercises substantial powers of management over the affairs of the company, yet such powers are exercised under the supervision and control of the board as a whole and as per the provisions of the Companies Act. Hence, it would not be fair to suggest that as soon as a managing director is appointed in a company, the powers of all other directors become ineffective. Furthermore, even though section 5 of the Companies Act, 1956 defines who are the 'officers in default', yet it does not exclude the outside independent directors from the purview of the 'officers in default'. Even a single shareholder can initiate proceedings to bring to book the directors of an erring company under section 621 of the Companies Act, 1956. This is a powerful tool in the hands of shareholders, which is however not frequently used by Indian shareholders.[16]

Further, the shareholders can level allegations that the independent directors utterly failed in their duty to protect the interests of the investors because the independent directors did not exercise the powers available under section 292A of the Companies Act, 1956 which mandates for the appointment of an audit committee consisting of outside directors.[17] The independent directors of Satyam Computer Services will have to explain whether they sought outside professional advice and help in cross-checking the veracity of the accounting statements, financial figures and bank documents submitted by the promoter's group while getting the accounts of the company approved by the board.

Role of Auditors

In his confessional letter, Mr Raju stated that the actual cash ($64.2 million) was only 5 per cent of what was shown ($1063 million) on the balance sheet. There was an understated liability of $246 million, beside non-existent accrued income of $75.2 million. In the last quarter, that ended on 30 September 2008, the actual profit margin was $12.2 million, while it was shown on the audited statement as $129.8 million. There were several items which were fake. SFIO investigations have unearthed another lie, in that $146.6 million shown in the records of Satyam Computer Services as existing in the New York branch of the Bank of Baroda proved to be non-existent. SFIO investigations also found defaults

in dividend payouts. It has been reported that the company has not followed the provisions under section 205 of the Indian Companies Act, 1956
which envisages dividends to be paid only out of profits. In fact the fraud
revealed may be even more serious than what has been confessed to; only
future investigation can paint the real picture.

By any yardstick, auditors and independent directors are men of emi-
nence and learning who should be independent. Yet this fiasco took place
on their watch. One can well imagine the Securities and Exchange Board
of India and the US Securities Exchange Commission wondering why a
board so exalted could not protect the interests of minority shareholders.
The biggest question raised is over the auditors. How one can rely on a
certified financial statement if it can be falsified to such an extent? It is a
failure of corporate governance system at each level, within the company
and outside the company.[18]

It is a matter of grave concern that top global auditing firm PricewaterhouseCoopers (PWC) signed off Satyam's financial statements for
a couple of years. PWC, which is the largest of the Big Four, had also
suffered a similar dent in its reputation in Japan three years earlier where
it was forced to shut down its affiliate after an alleged involvement with
scandal-hit stockbroking firm Niko-Cordial. The role of the auditors in
the Satyam fraud has a striking similarity with the Parmalat fraud. Only
recently, fraud amounting to 18 billion euros brought down Parmalat, the
Italian dairy giant, and ruined investors across the globe. Such an enormous fraud, some would assume, would need to be highly complex and
fully developed in planning as well as execution. However, as Parmalat
executives began to cooperate in the investigation, it was uncovered
how rudimentary their fraud was, despite the scale at which it occurred.
Parmalat, under the direction of Fausto Tonna, forged documents and
created fake transactions that any reasonable auditor should have been
able to uncover. Grant Thornton, the auditors in charge of auditing
Parmalat's financial statements, were seemingly asleep at the wheel in
confirming account amounts as well as their existence. The firm did not
confirm the forged documents with outside third parties such as banks
and other creditors. It also made questionable ethical decisions by auditing Bonlat, a spin-off of Parmalat SpA. Again questions were raised at
Grant Thornton. How did it manage to increase the auditing activities to
more than double despite losing the head auditor status? Also, by relying
on Parmalat to mail confirmation letters to various creditors, Grant
Thornton could have been directly involved in the fraud. Some believe
that Grant Thornton could have tipped the company off as to what con-
firmations were being mailed and when they were being sent out. Perhaps
the most surprising part of the fraud, which was never questioned by the

auditors, was the fake sales transactions which were so incredibly large that such a figure would be impossible.

After the Satyam fraud, auditors have come under scrutiny in other Indian companies.[19] There has been a demand from the Indian Institutional Investors for rotation of auditors to make sure that they do not act in collusion with company promoters.[20] The ICAI has recommended the recasting of financial statements to accommodate auditors' objections to any figure in the financial accounts, which is in conformity with global standards.

IMPLICATIONS

Sanctions and Enforcement

While the various regulations provide for sanctions for corporate governance violations, sometimes these are inadequate and should be adjusted. This is especially true for monetary sanctions. As regards the enforcing of the sanctions, the amended SEBI Act gives it certain pre-emptive powers, such as attachment of bank accounts. The MCA, SEBI and the stock exchanges share jurisdiction over listed companies. This creates a potential for regulatory arbitrage and weakens enforcement. An in-depth revision of the three-tiered regulatory system would reveal whether changes in the respective roles and responsibilities of the involved institutions and their supervisory functions are in order. Tables 4.1 and 4.2 show the nature of investigations taken up and completed by SEBI from 1996–97 to 2007–08.

Institutional Investors

Institutional investors could become important forces to monitor insiders and play a disciplining role in the governance of corporations. The behaviour of institutional investors has traditionally been characterized by apathy with respect to voting. The asset managers owned by financial institutions that maintain credit and advisory relationships with portfolio companies tend to feel uncomfortable when exercising their voting rights, especially against management. Policy-makers could play a role in encouraging institutional investors who act as fiduciaries to attend shareholder meetings and vote. This might encourage shareholder activism across the board, which is an important engine of change in corporate governance reform. Increasingly, international good practice suggests that it is preferable for institutional investors to nominate independent directors to the boards of their portfolio companies, rather than

Table 4.1   Nature of investigations taken up by SEBI

Particulars	1996–97	1997–98	1998–99	2000–01	2001–02	2002–03	2003–04	2004–05	2005–06	2006–07	2007–08	2008–09 P
Market manipulation and price rigging	67	29	40	47	47	86	95	96	110	137	95	12
Issue related manipulation*	35	15	4	2	5	1	2	2	2	3	0	0
Insider trading	4	5	4	3	6	16	13	14	7	6	18	7
Takeovers	3	3	6	1	1	1	9	2	1	4	2	2
Miscellaneous	13	1	1	3	9	7	6	7	10	15	5	4
Total	122	53	55	56	68	111	125	121	130	165	120	35

Notes:
P: Provisional
* Issue related manipulation includes fake stock invests

Source: SEBI.

Table 4.2   Nature of investigations completed by SEBI

Particulars	1996–97	1997–98	1998–99	2000–01	2001–02	2002–03	2003–04	2004–05	2005–06	2006–07	2007–08	2008–09 P
1	2	3	4	5	6	7	8	9	10	11	12	13
Market manipulation and price rigging	–	–	31	37	27	11	72	122	148	62	77	115
Issue related manipulation*	–	–	16	8	8	0	8	3	2	1	4	3
Insider trading	–	–	4	5	4	6	14	9	10	8	10	28
Takeovers	–	–	6	–	3	1	7	3	2	3	3	2
Miscellaneous	–	–	3	7	4	3	5	15	17	7	8	21
Total	–	–	60	57	46	21	106	152	179	81	102	169

Source: SEBI.

awarding the position to current or retired employees who may offer little value added and are more likely to be subject to conflicts of interest in the exercise of their fiduciary duties (ROSC, 2004). In the aftermath of the Satyam scam, the Life Insurance Corporation (LIC) and General Insurance Corporation of India (GIC), two prominent Indian institutional investors, have demanded rotation of auditors to make sure that they do not act in collusion with company promoters. This development may force the regulators to adopt this provision for the avoidance of such frauds in future.

Inadequate Information and Reporting Pattern

While analysing the annual reports of the companies, it has been found that they do not provide adequate information (Centre for Monitoring Indian Economy – CMIE, various issues). It is important that annual reports provide information on the acquisition of a substantial interest in another undertaking, or acquisition of a substantial interest in the reporting company by another group or company. If the acquirer happens to be an existing shareholder of the company, then total shareholding after the new acquisition should also be reported. Although SEBI has a Takeover Code, a large number of mergers and acquisitions deals do not come under its purview because they are not listed on stock exchanges, the acquisition involves the majority shareholder, or acquisition is a result of an overseas merger and acquisition between the foreign parents of the companies concerned. There is need for a more systematic and purposive transparent reporting on merger and acquisition activity in the Indian corporate sector (Kar, 2006).

The present format for reporting the shareholding pattern needs to be modified to reflect the ownership and control characteristics better. Given the level of aggregation and classification it is not possible from the present format to identify controlling interests and their stake in the risk capital. Shareholding of controlling interests should be identified separately in each of these categories, namely, foreign shareholding, intercorporate investments and the top 50 individual shareholders. It could be usefully done in the form of a short table, which gives the controlling and non-controlling interests in each of these categories. This is in addition to what is presently being reported under the shareholding of directors and their relatives. The foreign shareholding should be reported separately for foreign collaborators, foreign nationals, holders of American depository receipts (ADRs) and global depository receipts (GDRs), foreign institutional investors, non-resident Indians, and so on (Kar, 2006).

Insider Trading and Mergers and Acquisitions

In a comprehensive study of Indian mergers and acquisitions, it has been observed in the majority of the cases that there is a heavy build-up of share prices during the period leading to mergers and acquisitions announcements. This suggests the existence of insider trading on Indian stock exchanges (Kar, 2003). Here, SEBI needs to be more proactive as few persons privy to such prior information ought to benefit from such price-sensitive information. In all these years when SEBI has initiated probes of insider trading, very often it has been only after media outcry raised the issue. In the recent past the regulators have managed to catch some offenders. However, they got away with punishment not commensurate with their crime.

The Birla Committee has agreed on making it mandatory for some key management personnel in a company to disclose their sales and purchases of stocks. The committee is also of the view that companies must disclose only public information to analysts and researchers. But it is not clear whether the information gathered as a part due diligence before the merger and acquisition deal is struck will be treated as price-sensitive information and the deal would amount to insider trading.

TOWARDS A SUSTAINABLE SOLUTION

The spirit of corporate governance needs to be infused at the organisational level in the proper perspective going beyond mere legal compliance. This is imperative because corporate enterprises do not operate in an economy in isolation, and will have to coexist harmoniously with the natural and social environment. Corporate enterprises through responsible governance can create value for society, beyond the market, in innovative ways that are sustainable and inclusive. They may not be able to accomplish all this entirely on their own. They will have to forge public–private and community partnerships and craft their governance models that synergize shareholder value creation with the superordinate goal of ensuring inclusive and sustainable growth. However, to maintain a balance between profit and good governance practices for sustainable growth, corporate India needs to brace up for the following challenges.

Given that 75 per cent of population reside in India's rural areas, future sustainable growth will depend critically on finding solutions to rural India's unique problems. The lack of opportunities in rural India has also led to a huge migration of the workforce to urban areas, further adding to an already overstressed infrastructure. In its wake, it has also created a large mass of urban poor. It is therefore the responsibility of corporate India to create productive capacity and employment in rural areas, so that this segment is able to live a life of dignity in their home district and can be spared the pains of displacement.

Farmers, by and large, are at the mercy of the vagaries of nature and are worst affected by floods and droughts. Future climate change will impact upon them even further. Agriculture extension services are poor and the farmer has little or no access to know-how or technology.

Thus, corporate India needs to intervene and include the above-mentioned issues in its business models for a lasting solution. While there is no single 'one-size-fits-all' solution, the experiences of some of the established business houses of India, such as Tata, Birla, Infosys and ITC, give the impression that it is possible to achieve this synergy through innovative strategies and governance practices with a deep commitment that goes beyond the market.

DIMENSIONS FOR FURTHER RESEARCH

Researchers have tried to empirically investigate some aspects of the regulatory framework vis-à-vis company performance. Some researchers in the USA have looked for direct evidence of a link between board composition in terms of independence, and corporate performance. They have studied the correlation between the independent directors and the firms' performance as reflected by the accounting numbers. Baysinger and Butler (1985) and Hambrick and Jackson (2000) found evidence for the proportion of independent non-executive directors to be positively correlated with the accounting measure of performance. On the other hand, studies by Klein (1998), Bhagat and Black (1997) and Hermalin and Weisbach (1991) have found that a high proportion of independent directors does not predict a better future accounting performance. Using accounting measures, Agrawal and Knoeber (1999) found a negative relationship between board independence and firm performance. Hermalin and Weisbach (1991) and Bhagat and Black (2000) have used Tobin's Q as a performance measure, on the ground that it reflects the 'value added' of intangible factors such as governance (Yermack, 1996), and found that there is no noticeable relation between the proportion of outside directors and Q. The study by Lawrence and Stapledon (1999) produced no consistent evidence that the independent directors either add or destroy value, where corporate performance was assessed using accounting and share-price measures. Hermalin and Weisbach (1988) found that the proportion of independent directors tended to increase when a company performed poorly. Further,

Jensen (1993) opines that large boards can be less effective than small boards. His argument is based on the fact that when boards get beyond seven or eight people, they are less likely to function effectively and are easier for the CEO to control. A similar view is advocated by Lipton and Lorsch (1992) who state that the norms of behaviour in most boardrooms are dysfunctional because directors rarely criticize the policies of the top managers. In fact, they have recommended limiting the membership of boards to ten, with a preferred size of eight or nine. In fact, they suggest that even if board capacities for monitoring increase with the board size, the benefits are outweighed by such costs as slower decision-making, less candid discussions of managerial performance, and biases against risk taking. The inverse relationship between board size and performance has been reported by Yermack (1996), Eisenberg et al. (1998), Mak and Yuanto (2003) and Andres et al. (2005). However, Dalton et al. (1998), came up with contrary results.

Weirner and Pape (1999) have found that the system of corporate governance in a particular country is context-specific and is a framework of legal, institutional and cultural factors shaping the patterns of influence that stakeholders exert on managerial decision-making. Further, the board leadership structure outside of the US might be more varied and, hence, may have a different relationship with firm performance. This phenomenon is particularly true for transition economies experimenting with the Western forms of governance and market mechanisms. Nonetheless, firms outside of the USA are composed of different individuals and have different institutional expectations than those of the US, and this institutional context may lead to a different relationship with firm performance. There are very few research studies available covering the various aspects of effectiveness of the corporate governance mechanism in India. In a study of Indian companies, Garg (2007) has found mixed evidence that independent directors add value and improve the performance of the firm; at the same time there was no evidence that they destroy value. Further, Verma (1997) opines that there is no reason to expect the Anglo-American models of corporate governance to work in the Indian context. In fact, India had a unique system of managing agency in force for a long period of time before it was finally abolished (Kar, 2006).

This chapter has indicated the following key research dimensions on these aspects for further empirical investigation. Balasubramanian (2005) documents that Indian ancient texts have laid down sound principles of governance, which may be very relevant to modern-day sustainable corporate governance practices. There is a need for empirical investigation of this aspect to establish adherence by Indian companies.

The recommendations of the Confederation of Indian Industry Code on Corporate Governance, the Birla Committee, the Naresh Chandra Committee and the N.R. Narayana Murthy Committee are in favour of majority-independent boards, while the J.J. Irani Committee has recommended 33 per cent independence which can also vary with the size and the type of company. Some of the recommendations have been implemented in the recent past and there is a need to empirically probe whether board composition has any effect on a firm's performance.

Further, there is a need to investigate whether the cross-board phenomenon is affecting the performance of independent directors, which may be useful in framing the regulating guidelines on corporate governance while defining the criteria for a person to be eligible for appointment as an independent director.

Another area having wide implications is the compliance of various regulatory norms for corporate governance. A comprehensive study needs to be undertaken to analyse the degree of compliance of Indian corporate enterprises vis-à-vis their performance, including fulfilment of societal obligations.

In the post-Satyam scenario, there have been reports of large-scale resignations of independent directors. This is not to suggest that there is something problematic in every Indian company where independent directors have resigned in the past. That would be too rash a conclusion to draw. It could also be the fear of potential liability. Independent directors are often fearful about this issue for two reasons: (1) they are not involved in the day-to-day decision-making of the company although they may bear some responsibility for the actions of management; and (2) there are myriad directions from which liability could strike since directors are responsible (subject to exceptions) for violation of various statutes by companies. Literature on corporate governance has found that the risk of liability for independent directors is far lower than what commentators and directors themselves believe. It is reported that, even in the US, there were only a handful of cases where directors in fact had to make payments (and these include the high-profile Enron and WorldCom settlements). However, independent directors are indeed concerned, not about direct financial liability but about the time, cost, lost opportunity and reputational risk that accompany the mere initiation of legal action against them, even if that action does not succeed in the end. In the Indian context, the litigation process tends to be prolonged and cumbersome, with closure on the issue being unlikely within a short period of time. Hence, there is a need for a research study to find the rate of conviction and liability of independent directors for transgression of corporate governance practices.

Shareholders can bring to task the erring company management under various regulatory provisions. However, Indian shareholders are not exercising this option possibly because of the time, cost constraints and the level of confidence reposed on the regulatory bodies such as the Ministry of Corporate affairs (MCA) and SEBI which needs to be empirically investigated.

CONCLUSION

In the present business environment, corporate enterprises require greater autonomy of operations and opportunity for self-regulation to tide them over the financial crisis. At the same time, there is an urgent need to bring about more transparency through better disclosure and greater responsibility on the part of company management. It is worth noting here that no regulatory mechanism can enforce good corporate governance unless the participants follow the spirit behind these regulations. Hence, it is expected of the corporate enterprises that they will inculcate the culture of corporate governance within their organisations, which should be reflected in their actions.

NOTES

1. Father of the Indian nation, Mahatma Gandhi, wrote: 'I am inviting those people who consider themselves as owners today to act as trustees, i.e., owners, not in their own right, but owners in the right of those whom they have exploited.' In the Harijan paper, his views on trusteeship of property were later documented to clarify: 'It does not recognize any right of private ownership of property except so far as it may be permitted by society for its own welfare', and 'under State-regulated trusteeship, an individual will not be free to hold or use his wealth for selfish satisfaction or in disregard of the interests of society.' He also wrote that 'for the present owners of wealth . . . they will be allowed to retain the stewardship of their possessions and to use their talent, to increase the wealth, not for their own sakes, but for the sake of the nation and, therefore, without exploitation' (*Harijan*, 1 February 1942, p. 20).

2. The Irani Committee has suggested a minimum of one-third of the total number of directors should be independent directors, which should be adequate for a company having significant public interest, irrespective of whether the chairman is independent or not. There should be no requirement for a subsidiary company to necessarily co-opt an independent director on its board. The Irani Committee highlighted the issue as to whether the independent directors should be a majority or one-third of the board.

3. The Company Law Board, a quasi-judicial authority (now replaced by the National Company Law Tribunal), is the enforcement arm under the Companies Act. Enforcement matters referred to the CLB range from violation of procedure in a change of the main objects clause of the Memorandum of Association of the company, to oppression and mismanagement of the company. Some of the sections under which the CLB has enforcement powers are: 17, 18/19, 58A(9), 79/80A, 113, 118, 141,144, 163167, 186, 196, 219, 304, 307, 614, 621A, 111, 269(7), 634A, 22A (SCRA), 235, 237, 247/248, 250, 397/398, 408, 409 of the Indian Companies Act, 1956.

4. SEBI is an autonomous body established under an Act of Parliament, to which it

submits annual reports. The Cabinet appoints SEBI's board. SEBI's budget comes from fees, levies and government grants. Its decisions are subject to independent judicial review. SEBI is operationally independent, but the government can issue directions in policy matters. Investor protection and regulation of the securities market is one of the key mandates for the securities market regulator, SEBI. SEBI enforces market regulation and investor protection related to its jurisdiction by drawing from this mandate and related legislations such as the SEBI Act, 1992; Securities Contracts (Regulation) Act, 1956 and from the delegated provisions contained in Sections 55 to 58, 59 to 84, 108, 109, 110,112, 113, 116, 117, 118, 119, 120, 121, 122, 206, 206A and 207 of the Companies Act 1956.

SEBI has powers to carry out routine inspections of market intermediaries to ensure compliance with prescribed standards. It also has investigation powers similar to that of a civil court, in terms of being able to summon persons and obtain information relevant to its enquiry. The enforcement powers of SEBI include issuance of directions, imposition of monetary penalties, cancellation of registration and even prosecution of market intermediaries. Further, SEBI has the following powers under SEBI Takeover Regulations, 1997 and Prohibition of Insider Trading Regulations, 1994 towards achieving best corporate governance practices.

Any person who, directly or indirectly, acquires or agrees to acquire shares or voting rights in the target company, or acquires or agrees to acquire control over the target company, either by himself or with any person acting in concert with the acquirer shall on crossing 5 per cent or 10 per cent or 14 per cent, 54 per cent and 74 per cent inform the target company and stock exchange within two days. Persons holding between 15 per cent and 55 per cent, to disclose purchase or sales aggregating to 2 per cent or more, within two days to the target company and the stock exchanges. Open offer to purchase at least 20 per cent of shares at offered/determined price if holding crosses 15 per cent. Ban on using inside information for personal gain by way of purchase or sale securities for those who have access of internal information.

5. Under Clause 49 of the listing agreement, the board should consist of directors at least half of whom should be non-executive, independent directors. There has to be a minimum of four board meetings in a year. There is a prescribed ceiling on a director's membership in committees (maximum ten), and a ceiling on chairmanship of committees (maximum five). The Audit Committee chairman should be an independent director.
6. The two Indian rating agencies are CRISIL and ICRA.
7. The ICAI prescribed the following Accounting Standards for a better corporate governance mechanism: AS-1 Disclosure of Accounting Policies, AS-14 Accounting for Amalgamations, AS-17 Segment Reporting, AS -18 Related Party Disclosures, AS-21 Consolidated Financial Statements, AS-23 Accounting for Investments in Associates in Consolidated Financial Statements, and AS-27 Financial Reporting of Interests in Joint Ventures.
8. Regulations (2002) of the Indian Income Tax Act, 1961 contained the mechanism to ensure that income arising out of international transactions between related parties (associated enterprises) is computed on an arm's-length basis. Authorizes the assessing officer to refer the process of determination of arm's-length price to the transfer pricing officer amounts to reassessment.
9. Under sec. 227 (3d) of the Companies Act regarding whether, in opinion of the auditor, the profit and loss account and balance-sheet comply with the accounting standards referred to in sub-section (3C) of section 211, he may qualify his report and note observations or comments of the auditors which have any adverse effect on the functioning of the company. Further, Clause (6) Part I, Second Schedule of the Chartered Accountants Act, 1949 provides that failure of an auditor to report a known material misstatement in the financial statements of a company, with which he is concerned in a professional capacity, shall be deemed to be 'professional misconduct' which would invite action against the auditor.

10. With the amendment of the Companies Act, 1956 through the subsequent amendment in 1999 and the specifying rule in 2006, Accounting Standards have now indirectly become integral parts of the Companies Act, which will provide statutory backing. It says that every company and its auditor shall comply with the Accounting Standards in the manner specified in the rules. The Accounting Standards shall be applied in the preparation of general-purpose financial statements.

11. Apart from prosecution which can be initiated against the erring company directors under the Companies Act, the SEBI Act and the Securities Contract (Regulations) for non-compliances, it is important to note here that some of the provisions of the Indian Penal Code (IPC) may be applied against the company and its directors. The main charges against the company are related to the falsification and fabrication of accounts. When corporate entities resort to such illegalities as cheating, breach of trust, falsification and fabrication of books of accounts, forgery and so on, the state investigating machinery is asked to investigate and prosecute the accused directors for the offences, which involve penal consequences involving jail terms. The relevant sections of the IPC which are applicable are: 120B, 406, 409, 420, 468, 471 and 477A.

12. The World Bank alleged theft of data, violation of terms of business and indulging in bribes. Satyam also faced several lawsuits, including one filed in April 2007 by Upaid in the Texas court against Satyam for alleged fraud, forgery, misrepresentation and breach of contract involving transfer of intellectual property rights issues arising from a project the firms jointly worked on, and it asked for compensation of $1 billion.

13. Under Clause 49 of the listing agreement, the composition of the board should consist of at least 50 per cent non-executive, independent directors. However, the Irani Committee has recommended at least one-third of the board to be independent directors.

14. The list of independent directors on the Satyam board were: M. Rammohan Rao, Dean, Indian School of Business; K.G. Palepu, Professor, Harvard Business School; V.S. Raju, Former Dean, IIT Delhi; Mangalam Srinivasam; T.R. Prasad; and Vinod K. Dham.

15. The lack of financial acumen on the Satyam board is glaring. The company admits in its August 2008 Form 20-F filing with the SEC: 'We do not have an individual serving on our Audit Committee as an "Audit Committee Financial Expert" as defined in applicable rules of the SEC. This is because our board of directors has determined that no individual audit committee member possesses all the attributes required by the definition "Audit Committee Financial Expert".'

16. It is widely believed that shareholders are not exercising this option possibly because of the time, cost constraints and the confidence placed in the regulatory bodies such as the MCA, SEBI and so on. However, there is a dearth of research studies to empirically establish the facts.

17. As per clause (7) of section 292A of the Companies Act, 1956, the audit committee shall have authority to investigate any matter in relation to the items specified in this section or referred to it by the board, and for this purpose shall have full access to information contained in the records of the company and shall have external professional advice, if necessary.

18. The quality of financial disclosure for listed companies is determined by the MCA, SEBI and ICAI. The ICAI lays down the parameters of Accounting and Auditing Standards which are materially in conformity with International Financial Reporting Standards (IFRS) and International Standards on Auditing (ISA). Under the Companies Act, 1956, the management must explain any deviations from the prescribed accounting standards in the financial statements. Qualified auditor opinions do not prompt automatic action from SEBI. Moreover, the ICAI can take disciplinary action against its members for wrong auditing practices.

19. The two audit partners of PWC who have signed off the financial statements of Satyam, S. Gopalakrishnan and Srinivas Talluri, have already been arrested by the authorities.

20. Under the current Indian framework, only banks are required to rotate auditors, appointed from the Reserve Bank of India (RBI) empanelled list of auditors.

REFERENCES

Agrawal, A. and C. Knoeber (1999), 'Firm performance and mechanisms to control agency problems between managers and shareholders', *Journal of Financial and Quantitative Analysis*, 31(3): 377–97.

Andres, P.D., V. Azofra and F. Lopez (2005), 'Corporate boards in OECD countries: size, composition, functioning and effectiveness', *Corporate Governance*, 13(2): 196–210.

Balasubramanian, N. (2005), 'Corporate law reforms in India: management and board governance: two and a half cheers to the Dr J.J. Irani Committee', *Chartered Secretary*, 35(7): 978–84.

Basant, R. (2000), 'Corporate response to economic reforms', *Economic and Political Weekly*, 35: 813–22.

Baysinger, B. and H. Butler (1985), 'Corporate governance and the board of directors: performance effects of changes in board composition', *Journal of Law, Economics, and Organizations*, 1: 101–24.

Bhagat, S. and B. Black (1997), 'Do independent directors matter?' Working Paper, Columbia University.

Bhagat, S. and B. Black (2000), 'Board independence and long-term firm performance', Working Paper, University of Colorado.

Cadbury Committee (1992), *Report of the Committee on Financial Aspects of Corporate Governance*, Financial Reporting Council.

CMIE, Economic Intelligence Service (various issues), *Monthly Review of the Indian Economy*, Mumbai.

Dalton, D.R., C.M. Daily, A.E. Ellstrand and J.L. Johnson (1998), 'Meta-analytic reviews of board composition, leadership structure and financial performance', *Strategic Management Journal*, 19: 269–90.

Das, N. (2000), 'A study of the corporate restructuring of Indian industries in the post new industrial policy regime: the issue of amalgamations and mergers', unpublished PhD thesis submitted to University of Calcutta.

Eisenberg, T., S. Sundgren and T.W. Wells (1998), 'Large board size and decreasing firm value in small firms', *Journal of Financial Economics*, 48: 35–54.

Garg, A.K. (2007), 'Independence of board size and independence on firm performance: a study of Indian Firms', *Vikalpa*, 32(3): 39–60.

Hambrick, D.C. and E.M. Jackson (2000), 'Ownership structure, boards and directors', paper presented at the August 2000 meeting of the Academy of Management, Toronto.

Hermalin, B. and M. Weisbach (1988), 'The determinants of board composition', *RAND Journal of Economics*, 19(4): 589–606.

Hermalin, B. and M. Weisbach (1991), 'The effects of board composition and direct incentives on firm performance', *Financial Management*, 20(4): 101–12.

Jensen, M. (1993), 'The modern industrial revolution, exit, and the failure of internal control systems', *Journal of Finance*, 48(3): 831–80.

Kar, R.N. (2003), 'Corporate governance issues in mergers and acquisitions', *Chartered Secretary*, 33(August), 1181–3.

Kar, R.N. (2006), *Mergers and Acquisitions of Enterprises: Indian and Global Experiences*, New Delhi: New Century Publications.

Klein, A. (1998), 'Firm performance and board committee structure', *Journal of Law and Economics*, 41: 275–99.

Kothari, L. (1967), *Industrial Combinations*, Allahabad: Chaitanya Publishing House.

Lipton, M. and J. Lorsch (1992), 'A modest proposal for improved corporate governance', *Business Lawyer*, **48**(1): 59–77.

Mak, Y.T. and K. Yuanto (2003), 'Size really matters: further evidence on the negative relationship between board size and firm value', National University of Singapore Business School Working Paper, Singapore.

Naresh Chandra Committee Report (2002), Department of Company Affairs, Government of India, New Delhi.

Roy, M. (1999), 'Mergers and takeovers: the Indian scene during the 1990s', in Amiya Kumar Bagchi (ed.), *Economy and Organization-Indian Institutions under the Neo Liberal Regime*, New Delhi: Sage Publications, pp. 317–49.

Sampath, K.R. (2006), *Law of Corporate Governance: Principles and Perspective*, Mumbai: Snow White Publications.

SEBI (1997a), *Justice Bhagwati Committee Report on Takeovers*, Mumbai.

SEBI (1997b), *Securities and Exchange Board of India (Substantial Acquisition of Shares and Takeovers) Regulations*.

Verma, J.R. (1997), 'Corporate governance in India: disciplining the dominant shareholder', *IIMB Management*, **9**(4): 5–18.

Weirner, J. and J.C. Pape (1999), 'A taxonomy of systems of corporate governance', *Corporate Governance: An International Review*, **7**(2): 152–66.

World Bank, The (2004), Report on the Observance of Standards and Codes (ROSC): Corporate Governance Country Assessment, India, p. 16.

Yermack, D. (1996), 'Higher market valuation of companies with a small board of directors', *Journal of Financial Economics*, **40**(2): 185–212.

Web References

Confederation of Indian Industry (CII) (1999), 'Desirable corporate governance: a code', available at http:// www.ciionline. org/services/68/default.asp? Page=Corporate%20 Governances%20.htm (accessed 15 June 2003).

Lawrence, J. and G. Stapledon (1999), 'Is board composition important? A study of listed Australian companies', available at http://papers.ssrn.com/sol3/papers.cfm? abstract_id=193528 (accessed 15 June 2003).

SEBI (1999), 'Report of the Kumar Mangalam Birla Committee on Corporate Governance', available at http://www.sebi.gov.in/cms/sebi_data/attachdocs/ (accessed 12 May 2003).

SEBI (2003), 'Report of the N.R. Narayana Murthy Committee on Corporate Governance', available at http://www.sebi.gov.in/sebiweb/home/list/4/38/36/0/ Committee-Reports/ (accessed 15 February 2008).

5. Codes of conduct and other multilateral control systems for multinationals: has the time come – again?

Tagi Sagafi-nejad

INTRODUCTION

This chapter traces the evolving relationship between multinational enterprises (MNEs) or transnational corporations (TNCs)[1] and governments from the 1970s – when a protracted exercise under United Nations (UN) auspices to establish a code of conduct for MNEs was ultimately abandoned, through the 1980s when an Organisation for Economic Co-operation and Development (OECD) initiative to establish a multilateral agreement on investments was likewise given up. At the dawn of the twenty-first century, and especially after the 2008 global financial crisis, the need for multilateral instruments to delineate boundaries and establish rules of engagement for MNEs and nation-states is once again apparent. What can be learned from earlier attempts to establish such rules? Has the time indeed come for promulgating some form of international accord concerning the entire gamut of nation-states' relations with these enterprises?

The relations between MNEs and nation-states have been sometimes smooth and sometimes stormy. This ebb and flow has characterized the relationship since World War II. Firms and states have a symbiotic relationship and are indispensable to one another's survival and prosperity. Yet, conflicts of interest occasionally pit them against each other. In this chapter, I analyze the relationship by comparing and contrasting two periods, namely the turbulent 1970s and the first decade of the third millennium. I draw parallels, and explore the extent to which lessons learned during the earlier period can help us better understand the more complex and nuanced international business environment of the twenty-first century.

THEN AND NOW

Then

The 1970s have been called the leaden years (Ricupero, 2004), characterized by tumult and upheaval. Things began to unravel when the United States (US) unilaterally abandoned the dollar–gold link and the fixed exchange rate, sometimes referred to as the 'Nixon shock'. Other events added to the turbulence: the Munich Olympics massacre (1972), the October War (1973), the ensuing energy crisis (1973–74), the Watergate Affair (1972–74), the Chilean affair (1972–74), the overthrow of Allende in Chile and his death on 11 September 1973, and the negative repercussions of the Vietnam War. A more detailed account of this decade can be found in Sagafi-nejad and Dunning (2008). In 1973 and 1974, the US Congress held hearings and the US Securities and Exchange Commission conducted investigations into questionable conduct by MNEs: market manipulation with respect to oil, interference in the political affairs of certain countries, and other types of highly objectionable, if not illegal, conduct.

This misconduct led to various policy responses at both national and international levels. Many developed and developing nations enacted legislation outlawing bribery and illicit payments abroad. For example, in 1976, the US enacted the Foreign Corrupt Practices Act (FCPA). At the international level, the United Nations established a Group of Eminent Persons (GEP) to study the role of MNEs in development and in international relations and to propose certain policy measures. The GEP recommended the establishment of a Commission at the UN to focus on the role of transnational corporations and a center at the UN in New York to study the role of transnational corporations. See Appendix 1 for a summary of the Commission's charge. Other international initiatives included efforts by the OECD to devise guidelines for MNEs. Unlike the UN Code, the OECD guidelines have survived and evolved to this day. This period has been more fully explored elsewhere.[2]

Now

During the 1990s and into the first decade of this century, we again saw signs of discord regarding MNEs reminiscent of the tumultuous 1970s. The Enron debacle early in the new century was one of the most egregious examples of corporate misbehavior and one of the most closely studied. This and similar scandals engendered US Congressional responses, culminating in the enactment of the Sarbanes–Oxley legislation. Other corporate misdeeds involved Paramalat and ImClone, and an $11 billion accounting fraud at WorldCom. For this, its chief executive Bernie Ebbers was given a 25-year prison sentence. Corporate executives such as Dennis Kozlowski, the chief executive officer (CEO) of Tyco, Walter Forbes of Cendant, and executives at AIG, KPMG and other corporations also earned fines and incarceration. Bernie Madoff's Ponzi scheme was one of the most notorious and egregious of all. A few of these incidents of misconduct are summarized in Appendices 2 and 3.

Yet despite public outrage and national legislation, corporate misdeeds did not cease. The most recent round of misconduct resulted in the passage of the Dodd–Frank Wall Street Reform and Consumer Protection Act of 2010. This law is intended to make Wall Street firms more accountable by mandating companies to follow a series of strict reporting requirements. These caused the ire of critics who considered them too stringent. In 2011 the Securities and Exchange Commission (SEC), the US agency charged with overseeing corporate misdeeds, was scolded for failing to enforce such mandates. See Appendix 3 for selected headlines pertaining to cases of corporate misconduct.

Positive signs included a short-lived rejuvenation of the SEC under William Donaldson, its first proactive head in years, and the Office of the New York Attorney General under Eliot Spitzer. Donaldson and Spitzer stand out as leaders of efforts to cleanse the corporate world of chicanery in their respective spheres of control, and they exemplify government oversight at its best. Around the time of the global financial crisis in 2008, the SEC seemed to redouble its efforts in pursuing corporate misconduct. Many executives were called to account, and some were sent to prison. However, these hopeful signs of a more stern government response did not last. Neither Donaldson nor Spitzer could tame the beast and they left, each for different reasons.

I contend that while strengthening unilateral national rules and rigorously enforcing them is critical, globalization requires a collective approach to remedying corporate and governmental misbehaviour. Indeed, good global corporate citizens would likely regard transparent and standardized rules of engagement implemented uniformly as beneficial to their own best interests. Many corporate CEOs contend that they are prepared to abide by rules as long as they are fairly devised and enforced.

Potential platforms for articulating and negotiating such multilateral rules of engagement for TNCs and nation-states include the United Nations, the International Monetary Fund, the Group of 20, the World Trade Organization (WTO), and the Organisation for Economic Co-operation and Development. Each of these institutions has its own particular advantages and potential with respect to the articulation of these rules.

THE UNITED NATIONS

The UN was an early participant in the debate and indeed served as such a platform. One of the mandates of the United Nations Centre on Transnational Corporations (UNCTC), established as a result of a recommendation by the Group of Eminent Persons, was to devise a code of conduct for TNCs. The Centre was successful in many ways; however, the Code became its Achilles heel and the cause of its eventual demise. This was perhaps because this exercise became entangled in the East–West rivalry of the Cold War period. The core of UNCTC's activity, begun in New York in 1974, was transferred to the United Nations Conference on Trade and Development (UNCTAD) in Geneva in 1993 where it continues its work on foreign direct investment and related topics.[3]

Some 30 years later, in 1999, the UN again turned its attention to the *problematique* of TNCs, when Secretary-General Kofi Annan invited leaders of the global economy – primarily corporate executives – to engage with the UN in a 'global compact', a partnership between nation-states and TNCs, with the UN as intermediary. The Global Compact calls on companies to embrace, support and implement within their sphere of influence a set of ten principles in the areas of human rights, labour, the environment, and anti-corruption. These principles, grounded in earlier UN declarations, are:

- Human Rights
  Principle 1: Businesses should support and respect the protection of internationally proclaimed human rights; and
  Principle 2: make sure that they are not complicit in human rights abuses.
- Labour Standards
  Principle 3: Businesses should uphold freedom of association and effective recognition of the right to collective bargaining;
  Principle 4: elimination of all forms of forced and compulsory labour;
  Principle 5: effective abolition of child labour;
  Principle 6: elimination of discrimination in employment and occupation.
- Environment
  Principle 7: Businesses should support a precautionary approach to environmental challenges;
  Principle 8: undertake initiatives to promote greater environmental responsibility;

Principle 9: encourage the development and diffusion of environmentally friendly technologies.
- Anti-Corruption
Principle 10: Businesses should work against all forms of corruption,
including extortion and bribery.[4]

Now in its thirteenth year, the Global Compact initiative seems to have
found some traction, as many TNCs have discovered that 'doing well by
doing good' can indeed be a viable strategy. As of early 2012, over 8000
corporations and other organizations from 135 countries had joined the
Compact. The UN Global Compact Office, staffed by a skeleton crew,
diligently continues to educate corporations on its mission, and participation in the Compact has steadily increased over the years.

Some organizations may have joined to improve their public image or
to benefit from their association with the United Nations in other ways,
sometimes referred to as 'blue-washing'. To combat this potential abuse,
the Global Compact office has established a set of reporting requirements
through which participating entities demonstrate their compliance with
the ten principles. Signatories to the Compact who have failed to report
their progress are delisted or 'grey-listed'[5] so that all are aware of their lack
of compliance. As of early 2012, some 1005 companies and non-business
participants from over 135 countries have been delisted for such failure
to report, for one reason or another, their implementation of these principles.[6] The UN Global Compact's use of this kind of 'moral suasion' and
'soft power' may be as far as an international organization can currently
go, but it does nevertheless sow the seeds for more stringent rules and
further refinements.

The debate as to whether global rules should have legal enforceability
has been explored elsewhere (Sagafi-nejad, 2004: 363–82) and indeed
forms the crux of the debate. It was, after all, the developing countries'
insistence on codes binding on TNCs, and Western countries' adamant
opposition to them, that doomed UN efforts at devising a code of conduct
for TNCs and nation-states in the 1970s and 1980s. That debate deserves
new attention in light of subsequent changes in circumstances:

- The Soviet Union and Cold War rivalry no longer exist.
- Many developing countries have embraced the basic tenets of free
markets and open economies.
- Several developing countries have joined the ranks of developed
countries. Mexico and South Korea were invited to join advanced
nations as members of the Organisation for Economic Co-operation
and Development, as were Brazil, India and China. Even the

new Russia finds it to be in its national interest to join the global economy as an active player rather than a spoiler.
- Globalization, despite its shortcomings, has become accepted by a large number of nations, albeit in varying degrees.

These changes in the global economic landscape, the excesses and crises, have taught nations that they cannot act unilaterally. This imperative became evident when the US invaded Iraq, China attempted to flaunt global trade rules, and countries filed grievances against each other under the WTO Dispute Settlement Mechanism. Nations, large and small, must come to embrace the idea that straightforward rules, evenly administered, can move them forward.

Although UNCTC's earlier attempt at devising a code of conduct for transnational corporations failed, its various drafts can be instructive to those who might wish to revisit the issue. The last complete TNC code of conduct draft was published in September 1986. This 75-page draft, including an analysis by UN staff, consisted of three chapters and three appendices and drew on relevant UN resolutions. Its first chapter articulated the need for such a code by stating that TNCs play a positive role as effective instruments of development, but that their 'pervasive role . . . in the world economy requires the formulation of guidelines for their conduct' (UNCTC, 1974: 1). Other drafts, which continued to be negotiated until the entire exercise was abandoned, delineated the Code's structure, and addressed objectives, definitions and substantive issues as well as topics of current relevance, including TNCs' conduct, disclosure of information and political activities. They also contained text on the rights and duties of host-country governments and TNCs, including matters of nationalization, compensation and jurisdiction. Although the code was abandoned in the 1980s, its drafting process was instructive and texts of these drafts can guide future deliberations on global rules (see Appendix 4).

Despite progress in areas such as double-taxation and intellectual property, many other relevant issues remain unresolved today, and their resolution is passed on from one forum to another. Most concrete success has been achieved in areas related to trade and covered under the World Trade Organization mandate. But even here, progress is at glacial pace: witness the stalled Doha Round, to be discussed below.

When parts of the UN Centre for Transnational Corporations (UNCTC) were merged into UNCTAD, work on foreign direct investment (FDI) and TNCs continued, as did the notion of establishing rules of engagement palatable to all stakeholders. However, this work at UNCTAD took a more moderate turn, consistent with transformations in the global political and economic relations that followed the fall of the Berlin Wall and the collapse of the Soviet Union. Its focus became centered on capacity building for developing countries, investment agreements, rules on accounting, and international investment policy monitoring. This work is reported in the seminal *World Investment Report (WIR)*, which has been published annually since 1991. The *WIR* and a recent addition, *Investment Policy Monitor*,[7] assist developing countries in their efforts to attract FDI and its benefits. With these activities, this enterprise continues to fulfill its mission of leadership on policy matters, data and information pertaining to FDI and TNCs.

INTERNATIONAL MONETARY FUND (IMF)

The International Monetary Fund, a pillar of the tripartite institutional mechanism established under the 1944 Breton Woods system, took on new significance with the 2008 global financial crisis, precipitated in part by unbridled corporate excess. The IMF had devised its own 'code of good practices on fiscal transparency' in response to these crises (IMF, 2007). Its code delineates rules concerning responsibilities, the open budget process, availability of public information and assurances of integrity. While primarily addressed to the public sector and governments, its overlap with the 1976 UN Code on TNCs is noteworthy.

Since this IMF mandate is limited to financial and exchange rate matters, the organization has spearheaded norms, guidelines and procedures to enhance financial transparency among members and prevent deviation from these guidelines and procedures. In response to the Asian crisis of the late 1990s, the IMF initiated a Standards and Codes Initiative, which it described as follows:

> The Standards and Codes Initiative ('Initiative') has been identified as one of several building blocks for the overhaul of the global financial architecture after the Asian crisis in the late 1990s. Twelve policy areas were selected as key for sound financial systems and a framework for Reports on the Observance of Standard and Codes (ROSCs) was established and has been implemented by the Bank and the Fund for about a decade.[8]

THE GROUP OF 20

The fact that the G-7 (which became G-7 plus Russia, and then G-8) has been eclipsed by the Group of 20 signals a new era in the global economy when more countries are demanding to participate. The G-20 meeting in Pittsburgh on September 25, 2009 was more than ceremonial because this larger grouping of nations increases the desirability and feasibility of devising new global rules and learning from and building on those already in place so as to rein in excesses that led to the 2008 global financial crisis. In the communiqué that reflected the work of this meeting, the G-20 leaders agreed to:

- Launch a framework that lays out policies to generate strong, sustainable and balanced global growth;
- Make sure the regulatory system for banks and other financial firms will rein in excesses that led to the crisis;
- Reform the global architecture to meet 21st century needs;
- Maintain openness and move toward greener, more sustainable growth. (G-20, 2009)

The G-20 is thus engaged in global standard setting, if not outright rule-making, as it gains more saliency as a forum for debate on international policy.

WORLD TRADE ORGANIZATION (WTO)

The World Trade Organization, another pillar of the Breton Woods system, has become a major regulator and arbiter of trade-related matters since it came into existence in 1994 to succeed the 1947 General Agreement on Tariffs and Trade (GATT). Unlike GATT, WTO was empowered beyond mere suasion to adjudicate trade disputes and issue rulings binding on its signatory members. Nation-states, while jealously guarding their national sovereignty, nevertheless ceded some to this organization in the belief that such empowerment would ultimately serve their own national interests.

The organization has shown its effectiveness and resilience by settling trade disputes between contending parties, promoting and enhancing free trade, and gradually expanding its domain to include matters related to intellectual property and foreign direct investment. The growth of its membership is testimony to its usefulness to nation-states. Its ability to enforce its decisions gives it the 'legal teeth' lacking in other multilateral accords.

THE ORGANISATION FOR ECONOMIC CO-OPERATION AND DEVELOPMENT (OECD)

A club of some 30 rich democracies, the OECD actively works on rules of engagement pertaining to issues such as FDI, intellectual property, tax havens and bribery. In 1976 it succeeded in devising voluntary Guidelines for Multinational Enterprises, subsequently revised in view of changing times (see www.OECD.org). The Guidelines, part of a broader investment instrument called 'The Declaration on International Investment and Multinational Enterprises', are a set of recommendations to multinational enterprises in major areas of business ethics: employment and industrial relations, human rights, environment, information disclosure, combating bribery, consumer interests, science and technology, competition, and taxation. By promulgating these guidelines, these governments have committed themselves to promoting them among multinational enterprises operating in or from their countries.

The OECD also worked for a time to establish a Multilateral Agreement of Investment (MAI). While this effort failed, it taught the organization and its member countries valuable lessons on rule-making, particularly in regard to FDI and TNCs. These lessons are applicable as the current investment declaration works its way through the ranks of the organization's membership.

The OECD Guidelines are subject to frequent review, and the 2011 round of revisions resulted in a 65-page document, containing ten chapters:

- Concepts and principles
- General policies
- Disclosure
- Employment and industrial relations
- Environment
- Combating bribery
- Consumer interests
- Science and technology
- Competition
- Taxation.

A critical difference between the OECD Guidelines and the failed UN Code of Conduct for TNCs is that the former consists of a voluntary set of recommendations, while the latter – at the insistence of some developing countries – was to be mandatory, a difference that probably proved fatal for the UN Code. Yet the voluntary nature of any guidelines or rules poses its own problems because these are unenforceable and those negatively impacted are left without remedy. This is not to say, however, that voluntary guidelines are useless; they still serve to shed light on the rights and responsibilities of all parties. Such an educational outcome is itself worthwhile.

CONCLUSION

The twenty-first-century global economy – in which TNCs are major forces and economies are increasingly intertwined – calls for a set of transparent and multilaterally devised rules of engagement. If the institutional experiences discussed in this chapter, and the recent global financial crisis that has preoccupied world policymakers, can teach any lesson, it is that the need for openly devised and transparently implemented rules is more urgent than ever. As many observers have pointed out, the excesses exemplified by the appendices to this chapter are manifestations of globalization run amok. Observers such as Dunning, Stiglitz and Krugman have cautioned against them and have advised on ways to 'make globalization good', to borrow from Dunning.[9]

In 2000, Oliver Williams edited *Global Codes of Conduct: An Idea Whose Time Has Come*, which contained a variety of codes promulgated under various private and public auspices including the Sullivan Principles and other consumer and industry codes and standards. Contributors to that volume agreed on the need for global controls although they could not agree on the means to achieve this. In a study of the effectiveness of laws against bribery abroad, Alvaro Cuervo-Cazurra (2008) concluded that, to be effective, such laws must be coordinated across nations.

The question posed by Williams in the title of his 2000 volume can be answered in the affirmative. Some pre-existing institutional architecture, language and text can be built upon in light of new experience. It is feasible to use such precedents set by the UN, IMF/World Bank, G-20, WTO, OECD and others as a foundation for a set of standards for firms and states. Such norms, fortified with enforceability, can pre-empt the corporate and national misconduct which were precipitating factors in the call for stricter rules.

NOTES

1. While acknowledging the subtle distinctions made in the literature concerning nomenclature, here these terms are used interchangeably. The UN parlance is TNCs. See Sagafi-nejad and John H. Dunning (2008), pp. 2–3.
2. See Sagafi-nejad and Dunning (2008), especially Chapters 3–5.
3. See Sagafi-nejad and Dunning (2008), Chapters 5–7.
4. The tenth principle was added after the 2001 UN anti-bribery convention was adopted. See UN Global Compact website: http://www.unglobalcompact.org
5. The term was first used in Sagafi-nejad and Dunning (2008), p. 197.
6. A list of over 1000 of these companies and others who had joined the Compact was reported by the organization on its website in 2009. See http://www.unglobalcompact.

org/docs/news_events/9.1_news_archives/2009_10_07/Delisted_List_10_07_2009.pdfUH, retrieved 11 October 2009.
7. See http://unctad.org/en/Pages/Publications/Investment-Policy-Monitor.aspx
8. See http://www.imf.org/external/standards/index.htm for more details on Standards and Codes, adopted by the IMF and the World Bank in the late 1990s and which continue to evolve.
9. Among prominent persons who have spoken on the need to set standards to pre-empt misdeeds that precipitate crises, one can name John H. Dunning (2003) and Joseph E. Stiglitz (2002, 2006). Both have advocated some form of control and a stronger emphasis on corporate social responsibility.

REFERENCES

Cuervo-Cazurra, A. (2008), 'The effectiveness of laws against bribery abroad', *Journal of International Business Studies*, **39**(4): 634–51.
Dunning, John H. (ed.) (2003), *Making Globalization Good*, Oxford: Oxford University Press.
Ricupero, R. (2004), 'Nine years at UNCTAD: a personal testimony', Preface to UNCTAD, *Beyond Conventional Wisdom in Development Policy: An Intellectual History of UNCTAD 1964–2004*, Geneva: United Nations Conference on Trade and Development, pp. ix–xxxiv.
Sagafi-nejad, Tagi (2005), 'Should global rules have legal teeth? Policing (WHO Framework Convention on Tobacco Control) vs. Good Citizenship (UN Global Compact)', *International Journal of Business*, **10**(4): 363–82.
Sagafi-nejad, Tagi and John H. Dunning (2008), *The United Nations and Transnational Corporations: From Code of Conduct to Global Compact*, Bloomington, IN: Indiana University Press, for UN Intellectual History Project.
Stiglitz, Joseph E. (2002), *Globalization and Discontents*: New York: Norton.
Stiglitz, Joseph E. (2006), *Making Globalization Work*, New York: Norton.
United Nations (1982), *Draft United Nations Code of Conduct on Transnational Corporations*, U.N. Doc. E.C.10/1982/6, 5 June 1982, para. 13, reproduced in 22 I.L.M. 192 (1893) at 195 (as revised by *Draft United Nations Code of Conduct on Transnational Corporations*, U.N. Doc. E/1983/17/Rev.1, para. 13, reproduced in 23 I.L.M. 626 (1984) at 628).
UNCTC (1974), *The United Nations Code of Conduct on Transnational Corporations*, UNCTC Current Studies, Series A, No. 4, New York.
UN ECOSOC (1974), United Nations Economic and Social Council Resolution 1913 (LVII) E/RES/1913 (LVII) of 5 December 1974.
Williams, O. (ed.) (2000), *Global Codes of Conduct: An Idea Whose Time Has Come*, Notre Dame, IN: University of Notre Dame Press.

Web References

G-20 (2009), 'Pittsburg summit leaders statement', available at http://www.g20. org/Documents/pittsburgh_summit_leaders_statement_250909.pdf (accessed 11 October 2009).
International Monetary Fund (IMF) (2007), 'Code of good practices on fiscal transparency', available at http://www.imf.org/external/np/pp/2007/eng/051507c.pdf (accessed 4 June 2012).

APPENDIX 5.1 THE COMMISSION ON TRANSNATIONAL CORPORATIONS (1973)

The Commission:

Serves as the central forum within the United Nations system for the comprehensive and in-depth consideration of issues relating to transnational corporations;

Promotes an exchange of views among governments, intergovernmental and non-governmental organizations, trade unions, businesses, consumers and other relevant groups through the arrangement, *inter alia*, of hearings and interviews;

Provides guidance to UNCTC on the provision of advisory services to interested governments and the promotion of technical co-operation activities;

Conducts inquiries on the activities of transnational corporations, makes studies, prepares reports and organizes panels for facilitating discussion among relevant groups;

Undertakes work which may assist the Economic and Social Council in evolving a set of recommendations which, taken together, would represent the basis for a code of conduct dealing with transnational corporations;

Undertakes work which may assist the Economic and Social Council in considering possible intergovernmental arrangements or agreements on specific aspects relating to transnational corporations with a view to studying the feasibility of formulating a code of conduct and, on the basis of a decision of the Council, to consolidating it into a general agreement at a future date;

Recommends to the Economic and Social Council the priorities and the programmes of work on transnational corporations to be carried out by the Centre.

Source: Excerpted from the United Nations Economic and Social Council Resolution 1913 (UN ECOSOC, 1974).

APPENDIX 5.2   CONVICTED EXECUTIVES: A PARTIAL LIST

ImClone	Executive Dr Samuel Waksal, Pleaded guilty to securities fraud and bank fraud.	*BBC World News.com.*, October 15, 2002
WorldCom CEO	Bernie Ebbers accused and convicted of $11 billion in accounting fraud. Sentenced to 25 years in prison.	*Business on MSNBC. com.*, July 13, 2005
WorldCom	CFO Scott Sullivan pleaded guilty to manipulating the company's earnings in 2004 and cooperated with the government's prosecution of Ebbers. In return, he received only a five-year jail sentence.	G. Farrell, *USA Today*, November 12, 2006
Tyco	Dennis Kozlowski, former CEO sentenced to 8 years 4 months to 25 years in prison for stealing hundreds of millions of dollars from Tyco.	*CBS News.com*, February 11, 2009
ENRON	Skilling sentenced to 24 years in prison for conspiracy and fraud. Kenneth Lay allegedly died in July of 2006 at the age of 64. (October 2006)	*CNNMoney.com*, S. Pasha and J. Seid, May 25, 2006
Cendant	Walter Forbes, chairman, convicted of conspiracy and making false statements; forced to sell homes.	*USA Today*, G. Ferrell, November 12, 2006
Insurance Executives	Five insurance executivess convicted of conspiring to create a sham transaction that would inflate A.I.G loss reserves by $500 million.	*npr.com*, February 26, 2008
Smart Online	Dennis Michael Nouri, former CEO of Smart Online and his brother Reza Eric Nouri were convicted of all nine charges they faced in a stock fraud.	*WRALtechwire*, July 2009
KPMG	Robert Pfaff, et al., convicted of tax-evasion.	C. Bray, *Wall Street Journal*, December 18, 2008

Telecom executives – HK	Two top former execs convicted in swindling more than 12 million HK dollars in consultancy fees from China Motion Telecom International.	*Hphilstar.com*, February 5, 2009
Ernst & Young	Executives convicted of selling illegal tax shelters to wealthy clients.	G. Glovin, *Bloomburg. com*, May 8, 2009
HealthSouth	Richard Scrushy convicted of fraud totalling $2.7 billion over a seven year period; ordered to pay shareholders $2.9 billion.	HJ. Healy, *New York Times*, June 18, 2009
Bernard Madoff	Largest and longest Ponzi scheme in history. Sentenced in 2009 to 150 years in prison.	D. Hernriquez, *corporatewatch.org*, June 29, 2009
Cassey, Japan	Toshihachi Abe, former president convicted of violating Japan's unfair competition prevention law, mislabelling vegetables imported from China and selling as locally produced.	*Japan Times*, July 2009
Weizhen Tang	Multi-million-dollar Ponzi scheme, Overseas Chinese Fund, discovered by Ontario Securities Commission & US SEC.	Toronto *Globe and Mail*, January 13, 2010
USB Trader KwekuAdoboli	USB Trader charged with fraud appears in London court.	*Wall Street Journal*, September 16, 2011

APPENDIX 5.3   HEADLINES

Selected newspaper accounts of corporate scandals and their fallouts in the early part of the twenty-first st century:

'SEC scolded over scandals', Andrew Akerman, *Wall Street Journal*, November 2, 2011.

'A Madoff son hangs himself on father's arrest anniversary', Diana B. Henrique and Al Baker, *New York Times*, December 12, 2010.

'Latin American banks in line for more crises?' *San Antonio Express-News*, Nov. 20, 2004.

'Shell is urged to include investors in review process', *Financial Times*, June 25, 2004.

'Spitzer files suit seeking millions of Grasso money', *Wall Street Journal*, May 25, 2004.

'Grasso lawsuit fallout hits other executives', *Wall Street Journal*, May 25, 2004.

'Senate panel report deplores Riggs dealings', *New York Times*, July 15, 2004

'Spitzer sets sights on the pay of Grasso, once a "great friend"', *Wall Street Journal*, January 8, 2004.

'Of milk and money: arrests multiply as investigators cast their net wider in Parmalat's unfolding $10bn fraud', *Financial Times*, January 9, 2004.

'Philip Morris offers UT $1bn', *Financial Times*, April 5, 2004.

'Pfizer to plead guilty, pay $430 million to settle case', *Laredo Morning Times Business Journal*, May 17, 2004.

'Citigroup's Jones denies blame', *Wall Street Journal*, November 19, 2004.

'SEC expands its sleuthing to full sectors', *Wall Street Journal*, March 18, 2004.

'Prosecution winds up to climax in Tyco saga', *Financial Times*, January 20, 2004.

'Enron probe faults Skilling, Lay', *Baltimore Sun*, November 25, 2003.

'Two years after Enron, system isn't fixed yet', *New York Times*, January 20, 2004.

'US Technologies CEO pleads innocent to cheating investors', *Baltimore Sun*, April 3, 2003.

'Tyco sues ex-chief financial officer, claiming he owes firm $400 million', *Baltimore Sun*, April 3, 2003.

'James Comey, US Attorney, goes after Quattrone (CS-First Boston, fined $300 million), First to face criminal charges and possible jail', *CBS News*, April 23, 2003.

'SEC moves to empower corporation audit panels', *Baltimore Sun*, April 2, 2003.

'AIG May Pay Up to $90 Million', *The Wall Street Journal*, November 24, 2003.

'Blacklist of 'dirty money' havens put on temporary hold', *Financial Times*, September 26, 2002.

'Vivendi to face criminal probe by US watchdogs', *Financial Times*, November 5, 2002.

'EMTV chiefs go on trial in Germany charged with fraud', *Financial Times*, November 5, 2002.

'Exxon, Royal Dutch, BP Are in Price-Fixing Probe', November 11, 2002.

Source: The author, from public sources.

APPENDIX 5.4   SAMPLE PAGE FROM UN DRAFT CODE ON TNCS

This page illustrates areas of agreement and dissent among various negotiating parties. Items in brackets, often in multiple brackets, are worthy of detailed analysis, and provide a window into the areas of global discord, through the prism of the Code negotiations. Source: United Nations (1982).

–30–

[The Code is open to adoption by all States and is applicable in all States where an entity of a transnational corporation conducts operations.]

[The Code is universally applicable to all States regardless of their political and economic systems and their level of development.]

3.   [This Code applies to all enterprises as defined in paragraph 1 (a) above.]

[To be placed in paragraph 1 (a).]

[4.   The provisions of the Code addressed to transnational corporations reflect good practice for all enterprises. They are not intended to introduce differences of conduct between transnational corporations and domestic enterprises. Wherever the provisions are relevant to both, transnational corporations and domestic enterprises should be subject to the same expectations in regard to their conduct.]

[To be deleted]*

[5.   Any reference in this Code to States, countries or Governments also includes regional groupings of States, to the extent that the provisions of this Code relate to matters within these groupings' own competence, with respect to such competence.]

[To be deleted]

ACTIVITIES OF TRANSNATIONAL CORPORATIONS

A.   General and political

Respect for national sovereignty and observance of domestic laws, regulations and administrative practices

6.   Transnational corporations should/shall respect the national sovereignty of the countries in which they operate and the right of each State to

exercise its [full permanent sovereignty] [in accordance with international law] [in accordance with agreements reached by the countries concerned on a bilateral and multilateral basis] over its natural resources [wealth and economic activities] within its territory.

7.   [Transnational corporations] [Entities of transnational corporations] [shall/should observe] [are subject to] the laws, regulations [jurisdiction] and [administrative practices] [explicitly declared administrative practices] of the countries in which they operate. [Entities of transnational corporations are subject to the jurisdiction of the countries in which they operate to the extent required by the national law of these countries.]

8.   Transnational corporations should/shall respect the right of each State to regulate and monitor accordingly the activities of their entities operating within its territory.

---

*   On the grounds, <u>inter alia,</u> that the text within the first pair of brackets goes beyond the mandate of the Intergovernmental Working Group on a Code of Conduct.

6. Appropriate technology movement

Sanjeeb Kakoty

Man's desire to classify and understand human history on the basis of technology has often led to the classification of the past into periods such as the Stone Age, the Bronze Age, the Iron Age, the Age of the Steam Engine, the Information Age and the like. Not surprisingly, an examination of human history classified under different periods reveals the preponderant influence of a particular technology in each age. In addition, the essential fabric of human life and social mores of the society are often found woven around the predominant technology extant during that time. But the question that arises is: what determines the technology? Does technology arise due to specific physical need of the community? Is the technology in turn shaped and influenced by the philosophy, religious beliefs and world view of the particular community? In this scenario, what kind of impact would imported technology have on the specific need mitigation of the community, as also on their philosophy of life? Would it be possible and feasible to have a uniform technology for all regions of the world, or would it be better for disparate regions and individual communities to develop technologies best suited to their specific needs?

Any discussion on these questions would necessarily lead one to the centre stage of the appropriate technology movement, which was essentially a reaction against the trend of supplanting the unique production techniques and multiple economic systems prevailing in different regions, and replacing those with a single dominant model of an industrial mono system. Probably this trend had within it the seeds of the modern-day process of globalisation; both as a product as well as an objective of economic policy.

In order to gain a holistic understanding of the appropriate technology movement one has to go back to the very beginnings of the industrial age when questions were raised about the logic behind machine-driven economic growth, which often came at the cost of human well-being. Neil Postman provides an interesting list of people who expressed strong reservations about rampant industrialisation. William Blake likened factories to dark satanic mills which stripped men of their souls; while Matthew

Arnold felt that the greatest menace faced by humankind was the trend of placing too much faith in machinery. Writers such as Ruskin, William Morris and Carlyle linked it to spiritual degradation while Balzac, Flaubert and Zola highlighted the spiritual emptiness it evoked. Perhaps it was left to Lord Byron to powerfully argue against the ills of the industrial system. In a letter sent to Lord Holland dated 25 February 1812, he pleads for the adoption of a holistic view on mechanisation as it carried with it the danger of rendering redundant the labour available, impoverishing the individual and choking the vitality of the community. He wrote:

> Surely, my Lord, however we may rejoice in any improvement in the arts which may be beneficial to mankind, we must not allow mankind to be sacrificed to improvements in mechanism. The maintenance and well-doing of the industrious poor is an object of greater consequence to the community than the enrichment of a few monopolists by any improvement in the implements of trade, which deprives the workman of his bread, and renders the labourer unworthy of his hire. (Postman, 2000: 175)

In India too, similar views were being voiced, albeit in a later age and in a different situation. Mahatma Gandhi, who had assumed leadership of the Indian Freedom Movement, was quick to realise that the basic essence of the colonial relationship between Britain and India reduced India to an exporter of raw materials and a consumer of British finished goods. He railed against this equation, not only because it was an injustice being done to the country but also because it was based upon the edifice of a mass factory production system that was essentially dehumanising. Moreover, the process also invariably served a death blow to the self-sufficient and ecologically sustainable village economy of India, something that had been the hallmark of India over the centuries. The factory system, he pointed out, was not only driving the traditional village artisan and craftsmen out of business but also unleashing havoc in the social life of the communities in the Indian subcontinent. Since India was under the political control of Great Britain, it was seen as the natural downside of foreign domination. Thus, Gandhi's views on the use of technology were tempered with the aspirations of the Indian freedom struggle and also displayed strong moralistic opinions and beliefs. A votary of truth and non-violence, Gandhi posited the question in a larger human context and said:

> I must confess that I do not draw a sharp line or any distinction between economics and ethics. Economics that hurt the moral well being of an individual or a nation are immoral and, therefore sinful. Thus, the economics that permit one country to prey upon another are immoral ... The end to be sought is human happiness combined with full mental and moral growth. I use the objective moral as synonymous with spiritual. This end can be achieved under

decentralisation. Centralisation as a system is inconsistent with a non-violent structure of society. (UNESCO, 1958: 124)

In other words, economic systems that created modes of mass production stood in contrast to basic logic and purpose of human life. Gandhi went on to highlight the innate contradiction of the industrial system by commenting that 'if all countries adopted the system of mass production there would not be a big enough market for their products. Mass production then must come to a stop'. He went on to scrutinise the use of machinery and said:

> Machinery has its place; it has come to stay. But it must not be allowed to displace necessary human labour. An improved plough is a good thing. But if by chance one man could plough up, by some mechanical invention of his, the whole of the land of India and control all the agricultural produce and if the millions had no other occupation, they would starve, and being idle, they would become dunces, as many have already become. There is hourly danger of many more being reduced to the unenviable state. (UNESCO, 1958: 124–5)

Without a doubt, strong voices were emerging regarding the use of technology. Interestingly, what was being emphasised, over and over again, was the need to ensure that the success of any production process be measured against the end socio-economic result of that activity. Yet, it was apparent that man's obsession with the means pushed all talk about the end to virtual oblivion. In fact, what the nineteenth century witnessed may well be termed as megatrends towards mechanisation and mass production. Everyone seemed to be seeking newer and better ways to produce goods better, faster, more efficiently and in ever greater numbers. Factories churned out goods that required markets to sell and raw materials to produce. Profits were the *sine qua non* of the new age and it became an accepted practice to talk of market-driven economies in which welfare economics figured as a mere appendum. The quest for profits also resulted in attempts to further fine-tune the production process. This was the beginning of the discipline of management. It is interesting that the arguments forwarded in the Interstate Commerce Commission in 1910 about increasing freight rates on the railways led to the counter-argument that both wage increases and lowering of freight rates were possible by using the principles of what was termed as the 'Taylor System' (Postman, 1993: 50–51). The basics of this system were initially developed by Frederick W. Taylor and subsequently replicated all over the world. This was the beginning of the discipline of 'Scientific Management', and soon almost all sectors of the economy sought to use its methods to improve efficiency.

Efficiency of the production process was measured against greater profits and profits were, in most cases, the only goal of business. The profit motive assuming primacy had far-reaching ramifications. Labour was reduced to a factor in the production process and in order to achieve greater efficiency (read: more profits) it was considered legitimate to ignore human and humane dimensions in the production process. Efficiency was also sought to be increased through the introduction of better technology and improved machines. As a result, there were a large number of inventions and innovations that greatly transformed the manu-facturing process. The need to innovate and invent seemed to assume a momentum of its own and scientific inventions certainly became a hall-mark of this century. This led Alfred North Whitehead to observe that the greatest invention of the nineteenth century was the idea of invention itself. Interestingly, the question of how to invent things became primary and 'why' receded to the background (Postman, 1993: 42). Large-scale mechanisation followed and the industrial society based on mass produc-tion seemed to become the guiding mantra of the new age.

The resultant industrialisation of society, the creation of huge urban centres and the emergence of man as one of the factors in the production process, be it as a unit of labour or as a consumer, became the hallmark of the new economic system. The only discordant note that refused to die down was the essential dehumanising nature of the new system that uprooted traditional communities, systems and basic human values. By the mid-twentieth century there were strong reactions. The 1960s, which is associated with youth, protest, civil liberty and rebellion, also brought to the fore fundamental questions about the economy. Doubts were being expressed as to whether societies should strive for perpetual eco-nomic growth through rapid industrialisation, often ignoring the human and environmental costs. Moreover, rampant consumerism was creating monocultures of the mind, whereby the philosophy of one-size-fits-all technology was advocated as the answer to the needs of all communities irrespective of their location in the world. This trend unleashed havoc on local communities, their environment and their economies. What resulted was a search for an alternative blueprint for progress. This model had to be based on the use of appropriate technology. Appropriate technology meant taking into cognisance local needs, conditions and aspirations. Interestingly, these considerations automatically brought with them the principle of sustainability.

Perhaps the most celebrated work on this issue was that of E.F. Schumacher (1993). His book *Small is Beautiful* is considered to be a classic. It not only inspired the appropriate technology movement, but also put across strong arguments in favour of small-scale, decentralised, environmentally sustainable enterprises. He differed from the argument that the mere introduction of technology which may lead to increased production should be likened to progress, as in the process it may create unemployment, socio-economic disparity and environmental degradation. Certainly these cannot be features that are indicative of progress, whereas technologies that are small scale, decentralised and not energy intensive could be used to improve a community's standard of living. Such technologies also have the potential of being powered through renewable sources and hence being environmentally sustainable. These were termed 'intermediate technologies' and in 1966 Schumacher founded the Intermediate Technology Development Group (ITDG) in London. Today, technology that is designed to address the specific needs of local communities is termed 'appropriate technology'. The philosophical basis for this movement was sought to be provided in what Schumacher has termed 'Buddhist economics'. It is essentially the symbiosis between spiritual values and economic progress which can be achieved through 'right living'.

The quest for answers as to what constitutes 'right living' and the need to historically situate the philosophy necessarily takes one to ancient India. In the history of ancient India, the period between 1000 and 500 BC is identified as the era where the use of iron became popular, especially in the Gangetic plains. The use of iron technology especially in agriculture and warfare radically transformed the society. From being a pastoral economy that was largely migratory in nature, it slowly became an agricultural society that presupposed a sedentary and settled way of life. Advanced weaponry led to emergence of stronger and larger kingdoms with a uniform administrative system and a similarity of tools and technology used. The technology of the iron ploughshares necessitated the use of animal traction power such as bullocks and buffaloes, which were often in short supply due to the extensive killing of animals for ritual and other purposes. But in order to sustain the new iron economy, easy availability of animals that would provide the required traction power to operate the new tools was essential. What resulted was a miracle: India witnessed a revolution in religious philosophy, and individual life habits. The popularisation of the philosophy of non-violence, vegetarianism and reverence for the cow are some of the practices that emerged during this time, and religions such as Buddhism and Jainism that arose in sixth-century India spread all over the world (Sharma, 1990: 117–31). This is one of those classic instances in history where the adoption of a particular technology brought about a complete transformation of the philosophy and world view of the society. In turn, this new philosophy exercised a preponderant influence on future technology movements in the Orient.

It is tempting to extrapolate this argument and try and understand the flow of events in China. China for long called itself the 'middle kingdom', or the only kingdom that was civilized and was surrounded by people who were in various stages of barbarism (Vinacke, 1994: 38). The concept of the middle kingdom is also often interpreted as the kingdom between heaven and the earth. While the gods lived in the heavens above and the barbarians lived below, the Chinese occupied the middle kingdom. From their lofty abode, the Chinese viewed all other countries with thinly veiled contempt and treated their technology with disdain. Not surprisingly, innovation and change in the Chinese situation was more in the form of incremental change rather than radical transformation. In this situation, it may also be interesting to dwell on the fact that though it is the Chinese who are credited with the discovery of both the mariner's compass and gunpowder, it was left to the European powers to use these for world conquest. What could possibly explain this?

Perhaps the answer can be found in the contrasting philosophical moorings of the East and the West. As in the orient, in the occident too man's attitude to technology and innovation was largely shaped by religion and philosophy. Lynn White refers to the fact that Christianity inherited from Judaism certain traditions which include the concept of time as being non-repetitive and linear, and the story of creation in which God created man in his own image and then placed him at the centre of creation. This bequeathed him the inherent right to exploit nature in whatever way he chooses. It is probably from this belief that modern man imbibed the philosophy regarding the conquest of nature. By extension, his technological innovations too were largely disruptive in character. Thus it should come as no surprise that:

> By the end of the fifteenth century the technological superiority of Europe was such that its small, mutually hostile nations could spill out over all the rest of the world, conquering, looting, and colonizing. The symbol of this technological superiority is the fact that Portugal, one of the weakest states of the Occident, was able to become, and to remain for a century, mistress of the East Indies. (White, 1974: 2)

This philosophy seems to stand in contrast to the Buddhist philosophy which enjoins a reverent and non-violent attitude to all sentient beings (Schumacher, 1993: 49) and the appreciation that all beings are a part of the same creation. In fact:

> Concern for the welfare of the natural world has been an important element throughout the history of Buddhism. Recognition that human beings are essentially dependent upon and interconnected with their environment has given rise to an instinctive respect for nature. Although Buddhists believe humans have

a unique opportunity to realize enlightenment, which other creatures do not, they have never believed humanity is superior to the rest of the natural world. (Batchelor and Brown, 1994: 12)

Often dubbed 'Buddhist economics', as by E.F. Schumacher (Schumacher, 1993: 44–51), this philosophy appears tailor-made for sustainability. Under this, work is viewed as an activity with three main functions: that of giving man an opportunity to develop his faculties, to overcome an ego-centric existence by creating conditions to work with others for a common goal, and bringing forth the goods and services required for an optimum existence.

Creating the conditions for optimum human existence is not quite an easy proposition. Apart from its obvious philosophical moorings it would also require a strong scientific base. This may be greatly achieved through requisite technology and innovation. For instance, the use of the stirrup in medieval Europe radically transformed warfare and probably laid the foundations of the later feudal set up (White, 1964: 1–3). As this type of innovation often renders existing knowledge obsolete it is termed 'competence destroying'. At the other end of the spectrum is innovation and technology that is incremental in nature. Since it involves better use of existing knowledge it is termed 'competence enhancing' (Afuah, 2009: 15). Maybe it is not a coincidence that most of the Asian countries that have been influenced by philosophies such as Buddhism have adopted more of the incremental road to technological progress. As it were, the Buddhist attitude to technology displays their attitude to life and life goals, of which the economy and the production process is an intrinsic part. From the Buddhist point of view, there are two types of mechanisation which are clearly distinguished: 'one that enhances a man's skill and power and one that turns the work of man over to a mechanical slave, leaving man in a position of having to serve the slave' (Schumacher, 1993: 46).

Apparently, the Buddhist standpoint would stand in opposition to the conventional conversations on sustainability and innovations which have been essentially built on two basic premises. The first is that sustainable development can be attributed to a direct spin-off from perceived sustainable production processes. Secondly, at the root of this sustainable production process would lie certain innovative ideas that are translated into innovative production practices.

What is seen as a basic flaw in this line of thinking is that a linear relationship is sought to be established between sustainability and the production process. Confining the concept of sustainability only to the realm of production has within it the inherent danger of confining the larger sustainability debate within the narrow domain of the production process.

This would deny the issue the larger societal forum that encompasses not just the sustainability of the production process and the markets, but also the sustainability of the society itself, represented, as it may be, by its economy, social and political system, its education, its philosophy and its culture.

The appropriate technology movement has to encompass within it the larger social and economic debate. Perhaps a classic example of the influence of philosophy in the realm of the appropriate technology movement emerges from the agricultural fields of India. A story reported in a news magazine (Bavadam, 2011: 15–28) makes fascinating reading. It details the story of a young farmer by the name of Dadaji Khobragade, and dates back to the early 1980s. He lived in a village called Nanded Fakir in the Chandrapur district of Maharashtra. A person with a keen sense of observation, he once chanced upon a few paddy stalks which seemed to be sporting an unusual variety of grain. He kept the seeds aside and planted them in a particular field. When the plants matured and the stalks were adorned with grains he clearly saw the difference. Apart from colour, the other obvious difference was the shape of the grain, and the yield per plant seemed to be considerably higher. He was tempted to try it on a larger scale, so in 1988 he planted 4 kilograms of the new seeds in a 10 foot square field and harvested a good 400 kg of paddy. Much to his delight, he discovered that the new variety of rice had a delectable taste. Though the new paddy was a variant of the Patel 3 variety that had been developed by Dr J.P. Patel of the JNKV Agricultural University, Jabalpur, the new strain seemed more advanced. Khobragade's experimental fields were boasting of yields ranging from 40–45 quintals to a hectare, had a recovery rate of 80 per cent, and to top it all the rice was tasty and aromatic. Khobragade found it an instant hit in the market, and on being asked for a name instantly christened it after his wrist watch, HMT. Thus was born the famous HMT variety of rice and Khobragade lost no time in introducing the new seeds to his farmer friends. HMT spread far and wide, and today is planted in no less than five states covering some 100 000 acres of land. Not only has it shown resistance to pests, but because of the thinness of its grain it is today included as a standard reference for thinness by the Protection of Plant Variety and Farmers, Right Authority (PPVFRA). Khobaragade did not stop here. He relentlessly pursued his dream of discovering yet newer varieties, and in the last 27 years or so has developed no less than eight new varieties named after his village and his grandchildren. Ironically, he has derived no commercial benefit for his work save the grain he has grown on his own. On the contrary, the high cost of treatment for his son who suffers from sickle cell anaemia has forced him to sell his land. Interestingly, his name has been nominated by the National

Innovation Foundation for listing in *Forbes Magazine* as one of the seven most powerful Indian rural entrepreneurs who not only developed new seeds but also freely shared the knowledge with other farmers. It is estimated that his seeds are being used by some 200000 farmers in mainly five states of central India.

The legacy of Khobragade is certainly not an isolated story of a rural Indian making a tangible social impact through innovation. Stories of common people in different parts of the country coming up with remarkable technological innovations are being documented by an organisation called the National Innovation Foundation. Set up under government patronage in the year 2000, with the objective of 'recognising, respecting and rewarding innovations and outstanding traditional knowledge at the grassroots' (NIF, 2009b: 2), the NIF has complemented and supplemented the work of a group of dedicated volunteers who are organised under the Honey Bee Network. The organisation was initially started with a handful of volunteers and leading from the front has been the paternal figure of Anil Gupta, a management teacher in the prestigious Indian Institute of Management, Ahmedabad. During the past two decades the work of these volunteers has been transformed into a movement that scouts, spawns and sustains the unaided creative and innovative urges in the unorganised sector of society. Khobragade has received the support of the NIF and such is the case for hundreds of innovators from all over the country. Interestingly, the process of scouting for and documenting innovation in a relatively backward part of the country, the north-east, has produced extraordinary results. Innovators and their work have been documented in some detail.[1]

After locating innovators and their innovations from remote areas, the challenge is to transform them to commercially viable propositions. One innovation that has attracted the attention of the press (Gani, 2011: 3) is a wind-operated pump. Designed by two brothers, Mohammad Mehatar Hussain and Mushtaq Ahmad, the wind pump was invented to help irrigate the 2 acres of paddy fields they owned in Sipajahar, Assam. Designed with locally available bamboo, tin sheets and discarded tyres, the basic design has undergone some modification with the help of the NIF. The water discharge from this pump is 1500 to 1600 litres per hour with a required wind speed of 8 to 10 km per hour. Such wind speed is a normal occurrence in most places of this state and is expected to be a boon for an electricity-deficient state like Assam. The model has been commercialised and costs between Rs 6000 and Rs 40000. Data are not available to show its market performance.

If the experience of the NIF is anything to go by, innovations and market dynamics do not exactly go hand in hand. Many brilliant innovations, be it the pomegranate de-seeder of Uddhab Bharali or Kanak Gogoi's innovation to generate electricity out of vehicles moving over rumble strips, are yet to be commercialised and marketed (NIF, 2009b: 20); innovation in marketing is probably what is required.

A recent initiative to market craft items certainly has the potential to transform the marketing dynamics of the rural artisan and craftsmen. It comes in the form of making available the benefits of the Internet and its unprecedented market reach to the artisans, many of whom are unlettered and live in remote villages. Their products and wares are displayed in cyberspace and prospective customers can also place orders online. The traditional artisans of India are being brought under one roof – or rather one web, on www.UnWrapIndia.com, a website dedicated to bringing these products from rural India to the urban consumer. Products are being sourced from non-governmental organisations (NGOs) supporting destitute women, blind children and rural artisans, and a host of others. Within the first few months of its existence, the huge potential of the market in India as well as abroad became apparent. With a product range of some 2800 products from 95 different suppliers across India which include rural artisans from places as far away as Manipur and Assam, the aim is to have a product range of at least 10000 products by June 2012 and a robust supply chain for these. Plans are also afoot to start an offline store after June 2012.

There is no doubt that these initiatives do augur well for the appropriate technology movement. Another development that has the potential to provide a tremendous boost to this is the growing momentum of the open source initiative. What essentially started with a move to open source the realms of the cyber world through software and data sharing has begun to envelope every other sphere. Now there is talk of 'open life ecology'. One name that stands out in realm of open sourcing hardware is that of Marcin Jakubowski. With a PhD in Fusion Physics, he went on to become a farmer in Missouri who started the initiative of Open Sourcing DIY (do it yourself) of what he termed a 'global village construction set'.

The emphasis of his initiative was to help people build their own low-cost tools with locally available material that would last a lifetime. Jakubowski feels that such an initiative would be sustainable and also prove that industrial productivity on a small scale was possible. He feels that it could be the answer that would ensure proper distribution of the means of production and put in place an ecologically sound supply chain. Underlining the potential of his initiative, Jakubowski comments that the knowledge of open source hardware could be transmitted through a simple CD. The simplicity of transmission belies the tremendous impact of the free exchange of knowledge. This probably prompts Jakubowski to term it the 'virtual blueprint for civilisation'.

These initiatives certainly have the capacity to help what David Korten calls the 'countless local microenvironment'. The diversity of cultures around the world is indicative of the vibrancy of each of these micro systems. The dynamism and creativity of micro systems needs to be recognised and respected. If they are allowed to flower according to their own genius, it would 'optimise the capture, sharing, use and storage of available energy and material resources both for itself and as its contribution to the needs of the larger system this optimisation is possible because an eco-system is local everywhere it touches Earth' (Korten, 2009: 108).

Though it may be tempting to depict the process of innovation as a smooth, well-regulated linear process, in reality that rarely is the case (Rosenberg, 2010: 173). Instead what is more apparent is the complex interplay of numerous components that make up the multiple strands of the innovation process. Interestingly, the different strands or factors often operate independently of each other. Probably this gives rise to the ecological variety that Korten talked about.

Obviously, the challenge facing a firm seeking successful innovation is largely commercial, though its technology is equally important. An anecdote may be cited the case of an invention made by the redoubtable Thomas Edison. It was a voting machine that would instantly tally the votes cast. Instead of meeting with success, Edison was told by a number of Congressmen that it was the last thing they wanted. This machine failed the market test and the journal entry of that day, made by Edison, spoke of a resolution never again to spend time on an invention until he was sure a sound market existed (Rosenberg, 2010: 176).

As customer acceptance and market demand largely determines the success of any innovation, appropriate technology should represent a healthy coalescence between invention and commercialisation (Afuah, 2009: 13). The ability of the firm to garner profits out of innovations rests in its ability to use the new knowledge and offer its product at a lower cost than its competitors or create a well-marked differentiated product that would command a premium price. This in time would emerge as the defined area of competence of that company and ensure its advantage over its competitors.

One essential imperative of modern business is the need for perpetual growth. 'Growth is important because companies create shareholder value through profitable growth' (Christensen and Raynor, 2003: 1). Current wisdom believes that it is through growth that companies create shareholder value. The need for constant growth often pushes companies to newer and often unchartered trajectories. It is estimated that merely:

> One company in ten is able to sustain the kind of growth that translates into an above average increase in shareholder returns over more than a few years. Too

often the very attempt to grow causes the entire corporation to crash. Pursuing growth the wrong way can be worse than no growth at all. (Christensen and Raynor, 2003: 1)

There is also the problem of the high proportion of what is termed failure or dropout concerning the formulation of a new idea to its eventual market launch. It is indicated that the proportion of ideas falling by the wayside ranges from about 30 per cent to as high as 95 per cent (Conway and Steward, 2011: 278). With such a high percentage of failure in market convertibility of ideas, there is little doubt that firms may appear wary of committing huge funds to generate innovations. This trend also seems to supplement the apology of organisational behaviour exponents that the stimulus–response theories are overly mechanistic and reductive in nature (Fincham and Rhodes, 2010: 40).

To make matters worse, 'since most companies do not know how to measure their technological health, they measure their economic health. The trouble is that economic health is a result of many things that essentially are independent of the underlying technological health of the firm' (Foster, 1986: 153). In order to ensure flow of innovation into their process, companies often resort to seeking help from outside. It has been commented that the biggest single trend seems to be the growing acknowledgement of innovation as a centrepiece of corporate strategies and initiatives (Kelley and Littman, 2008: 3).

That the appropriate technology movement has to go beyond the confining limits of innovation, there is no doubt. If one were to take the classic example of solar energy, the issue becomes very apparent. Solar energy provides the curious spectacle of a situation whereby in spite of billions of dollars spent on research and use, it is nowhere near profitable commercialisation. Probably the approach of comparing solar energy with conventional energy and putting it in competition against the energy being supplied by the established grids dooms it to failure.

About two-thirds of the world's population has access to electric power transmitted from central generating stations. In advanced economies this power is available almost all the time, is a very cost-effective means of getting work done, and is available essentially 24 hours per day, cloudy and sunny weather alike. This is a tough standard for solar energy to compete against (Christensen and Raynor, 2003: 109). Interestingly, the scenario is reversed the moment solar energy is offered to the hitherto untapped segment that has no access to conventional electricity:

[For] the two billion people in south Asia and Africa who have no access to conventionally generated electricity, the prospects of solar energy might look quite different. The standard of comparison for those potential customers is no electricity at all. Their homes are not filled with power hungry appliances, either, so it would be a vast improvement over the present state of affairs for these customers if they could store enough energy during daylight to power an electric light at night. (Christensen and Raynor, 2003: 109)

Interestingly, it is being increasingly realised that structuring and directing the innovation process may greatly increase returns for the company. The starting point of this would obviously include understanding the environment. This is termed 'scanning the environment' and is defined as the systematic methods used by a company to monitor and forecast those forces that are external to and not under the direct control of the organisation. Nicolau (2007) states: 'Given that the environmental factors and actors can influence the future of the company, top managers may envision their effects, to take advantage of opportunities and defend from threats, and to measure their impact on performance'. He also stresses that 'the extent that a firm's ability to adapt to its outside environment depends on knowing and interpreting the external changes that are taking place, environmental scanning constitutes a primary mode of organizational learning' (Nicolau, 2007: 143).

What kind of organisational learning advantage do Indian firms display when they succeed in cornering the major share of the outsourced Western market? It is said that India accounts for 60 per cent of the offshore white-collar jobs market. Its presence in the growing pool of high-value jobs leaving the USA – and, crucially not being created there – seems set to accelerate (Seshabalaya, 2005: 1).

The process of the relocation of white-collar technology jobs out of the West is a powerful undercurrent in today's globalising world economy. One study by the US consulting firm Forrester Research estimates that such a process could send 3.3 million jobs overseas by 2015. Another by Deloitte Consulting predicted in 2003 that Western financial companies alone would move a total of 2 million jobs over the next five years (Sheshabalaya, 2005: 7).

The issue of job outsourcing does seem to be an area of immense interest for the media and academics, not to mention politicians. As a case in point may be cited the incident involving the US state of Indiana which had decided that its work on upgrading the database of it jobless claims system would be outsourced to the Indian company TCS. Predictably, it ignited a huge debate. Ultimately, Indiana rescinded the contract. Apparently the move was aimed at ensuring greater opportunities for local companies and thereby helping the economy. On the emotive plane of appeasing public sentiment, the move cannot be faulted. But on the other side stand certain hard figures that are hard to ignore. For instance, the $15.4 million bid of

TCS was a whopping $8.1 million lower than that of its rivals. This figure is rather substantial when one considers the fact that the duration of the contract period was a mere four months. In simple terms, 65 Indiana jobs were extended for four months at a cost of $8.1 million (Seshabalaya, 2005: 34–5). Whether this type of response, which is becoming increasingly common, can be considered adequate in technological and economic terms, does not merit serious debate.

Probably there is more to be learnt by focussing attention on the role of cultural idiosyncrasies in the process of decision making. In this regard, it is important to understand a basic divergence between Indian and Western mindsets: while the essential content of Western culture is form, that of Indian culture is content. An interesting observation made way back in 1997 by Karen Elliot House, president of the Dow Jones company, is worth recounting. She observed that, unlike China, India was 'chaotic' on the surface but was 'stable' underneath. Probably, this order in chaos can be seen at work in the functioning of the Indian railway system. The British newspaper the *Sunday Times* discovered in September 1998 that India's 14 000 trains fare better on timing than those of Great Britain. The key element of Indian business seems to be in providing substantive content.

As a viable business model that draws heavily on content versus form debate may be cited the working of the famous *dabbawallas*. (People dealing with *dabbas*, from the Hindi word *dabba* or small container, referring to the complex system of lunch box delivery from homes to office that started in the city of Mumbai.) *Forbes* magazine reported that the 175 000 daily deliveries of lunchboxes on foot, bicycles and trains by Mumbai's redoubtable 'tiffinwallahs', amidst the city's dust, grime and seeming chaos, yields just 'one error in every eight million' deliveries. For enterprise software portal ERPWeb.Com, this is one of the handful of achievements worldwide to the futuristic Six Sigma quality standard (Sheshabalaya, 2005: 253).

Seeking further examples of unique Indian enterprises that are worthy of emulation, one may cite the examples of brand Amul and the Arvind eye hospitals. Undoubtedly, Amul provides the best example of the perfect convergence between the shareholder and stakeholder space and the creation of a business model that is absolutely unique. Amul is a business proposition that consists of millions of members of a co-operative who generate annual business in excess of $2 billion. 'Its daily milk procurement is approx 12 million lit (peak period) per day from 15 712 village milk cooperative societies, 17 member unions covering 24 districts, and 3 million milk producer members.'[2]

This milk co-operative Gujarat Cooperative Milk Marketing Federation

(GCMMF), more known by its brand name Amul, began its journey in 1946. It is credited with bringing about the White Revolution in India which transformed India from a milk deficient country to a major exporter. Today it is arguably the largest food brand in India and the world's largest pouched milk brand with an annual turnover of US$2.2 billion (2010–11). It is also the world's largest vegetarian cheese brand. Currently unions making up the GCMMF have 3.1 million producer members with a milk collection average of 9.10 million litres per day.[3] It is amazing how a once small co-operative society transformed India into becoming the largest producer of milk and milk products in the world, and the profits of this enterprise are shared by millions all over the country.

Probably the other great Indian story is that of the Arvind Eye Hospitals. What started as a dream of Dr Govindappa Venkataswamy, who in 1976 used his pension money to start a modest 11-bed eye hospital, is today the largest and most productive eye care facility in the world. Interestingly, Dr G. Venkataswamy modelled his hospital on the lines of the redoubtable McDonald's, when he sought to replicate a model that had succeeded in delivering consistency in quality and product in their numerous McDonald's outlets in diverse geographical areas. The Arvind hospital wanted to provide top-class eye care to patients, even in remote areas, irrespective of their ability to pay. Today the hospitals and outreach centres perform more than 200 000 operations each year, two-thirds of which are done absolutely free. The costs of those unable to pay are often covered by the paying category of patients. The sheer economies of scale at Arvind Hospitals are also overwhelming. Surgeons here perform over 2000 operations annually against the national average of 220. Research into lens and other eye implants has also produced spectacular results. In the early 1990s intraocular lenses had to be imported and these lenses cost US$200. This effectively put them out of reach of the common man. Arvind developed their own lenses which were priced at US$5, and today these are exported to 85 countries.[4]

There are numerous other examples that exemplify the Indian way of doing business. R.A. Mashelkar terms this phenomenon 'value for many' as opposed to 'value for money'. These trends straddle the broad spectrum of business endeavours, from co-operative initiatives to corporate ventures. The Tata decision to produce the world's cheapest car, the Tata Nano, was an interesting blend of business strategy and social altruism. If the move to make the cheapest car aimed to democratise private transport and was aimed at the social good, the fact that it captured a huge untapped market cannot be overlooked. One may also take the example of the cheap, mass-produced artificial limb called the Jaipur foot, being made available to thousands of amputees at an unimaginable cost. It is interesting to dwell on what propels these businesses: philanthropy? Profit? Or a mix of both?[5]

Perhaps the mantra that these hugely successful Indian businesses are offering to the world is the need to blend the two P's: philanthropy and profits. The same philosophy seems to influence governance in India and offers the unique spectacle of state intervention working for the common good.

A talking point has been the proactive role being played by the judiciary, from issues related to corruption in policy-making to pollution control, child labour and education. India's Supreme Court continues to set new global milestones in socially aware judicial activism – for example playing a direct role in New Delhi's success in converting the entire public transport system to compressed natural gas (CNG) propelled vehicles, on a scale unparallelled elsewhere. Its independence was hailed by Britain's Lord Chief Justice as a lesson for the United Kingdom (Sheshabalaya, 2005: 230–31).

Are these stories examples of appropriate technology at work? Are these expositions a part of what has been termed the 'bottom of the pyramid' model, made famous by C.K. Prahalad? There is ample reason to believe that whatever nomenclature is used, these are benchmarks that corporations all over would do well to incorporate. And as Nandan Nilekani, a well-known voice of corporate India, sums up: 'this is what is unique about the Indian growth story. A people driven transformation of a country holds a particular power; it is irreversible' (Nilekani, 2009: 475).

NOTES

1. For more details see http://www.nif.org.in/dwn_files/assam/PART-I%20ASSAM%20 BOOK.pdf (accessed on 23 May 2012).
2. The Amul official website, available at http://www.amul.com/m/organisation (accessed 23 May 2012).
3. The story of Amul is available at www.wikipedia.org/wiki/Amul (accessed 6 April 2012).
4. Arvind available on www.aravind.org (accessed 6 April 2012).
5. Tata Nano and Jaipur Foot, available at http://www.youtube.com/watch?v=z_ XchYY3bnU (accessed 6 April 2012).

REFERENCES

Afuah, A. (2009), *Innovation Management*, New York: Oxford University Press, Indian Edition.
Batchelor, M. and K. Brown (eds) (1994), *Buddhism and Ecology*, Delhi: Motilal Banarsidass.

Christensen, C.M. and M.E. Raynor (2003), *The Innovator's Solution: Creating And Sustaining Successful Growth*, Boston, MA: Harvard Business School Press.

Conway, S. and F. Steward (2011), *Managing and Shaping Innovation*, Oxford: Oxford University Press.

Fincham, R. and P.S. Rhodes (2010), *Principles of Organizational Behavior*, New York: Oxford University Press.

Foster, R. (1986), *Innovation: The Attackers Advantage*, New York: Summit Books.

Gani, A. (2011), 'Two Assam farmers invent wind operated pump', *Seven Sisters Post*, Guwahati, p. 3.

Kelley, T. and J. Littman (2008), *The Art of Innovation*, London: Profile Books.

Korten, D.C. (2009), *Agenda for a New Economy*, Tata, New Delhi: McGraw Hill.

Nicolau, J.L. (2007), 'Gaining strategic intelligence through the firm's market value: the hospitality industry', in Mark Xu (ed.), *Managing Strategic Intelligence: Techniques And Technologies*, Information Science Reference, Hershey, New York: IGI Global, p. 143.

NIF (2009a), 'Creative North East', Ahmedabad.

NIF (2009b), 'In support of grassroots innovation and traditional knowledge', NIF Update 2009, Ahmedabad.

Nilekani, N. (2009), *Imagining India Ideas for the New Century*, New Delhi: Penguin Books.

Postman, N. (1993), *Technopoly*, New York: Vintage Books.

Postman, N. (2000), *Building a Bridge to the 18th Century*, New York: Vintage Books.

Prahalad, C.K. (2005 [2008]), *The Fortune at the Bottom of the Pyramid: Eradicating Poverty through Profits*, Wharton, PA: Pearson Education, Wharton School Publishing, 4th impression, 2008.

Rosenberg, N. (2010), *Studies on Science and the Innovation Process*, Singapore: World Scientific Publishing.

Schumacher, E.F. (1993), *Small is Beautiful*, London: Abacus.

Sharma, R.S. (1990), *Material Culture and Social Formation in Ancient India*, Madras: Macmillan India.

Sheshabalaya, A. (2005), *Rising Elephant: The Growing Clash With India Over White Collar Jobs and Its Challenge to America and the World*, Delhi: Macmillan India Common Courage Press, Indian reprint.

Shiva, V. (1993), *Mono Cultures of the Mind*, London, UK and New York, USA: Zed Books.

UNESCO (1958), *All Men are Brothers, Life and thoughts of Mahatma Gandhi in his own Words*, Calcutta, India and Paris, France: Orient Longman.

Vinacke, H.M. (1994), *A History of the Far East in Modern Times*, New Delhi: Kalyani Publishers.

White, Jr. L. (1964), *Medieval Technology and Social Change*, Oxford: Oxford University Press.

Web References

Bavadam, L. (2011), 'Bitter harvest', *Frontline Magazine*, **28**(2) available at http://www.frontlineonnet.com/fl2802/stories/20110128280203700.htm (accessed 6 April 2012).

Jakubowski, M. (2012), 'Open-sourced blueprints for civilization', available at http://www.ted.com/talks/marcin_jakubowski.html (accessed 20 March 2012).

White, L. (1974), 'The historical roots of our ecologic crisis [with discussion of St Francis; reprint, 1967]', available at http://www.siena.edu/ellard/historical_roots_of_our_ecologic.htm (accessed 1st December 2011).

PART II

Strategic implications and assessment

7. Eco-social business in developing countries: the case for sustainable use of resources in unstable environments

Roland Bardy and Maurizio Massaro

INTRODUCTION: OBJECTIVES AND METHODOLOGY

Despite recent interest in the 'new' social enterprise movement, the issue of how to connect this successfully to ecological concerns has not yet found widespread interest. There is much development in the relations between business and social ventures: for example, there is an extensive history of nonprofits using fees for services and other revenue-raising techniques in order to supplement or complement their mission activities (Alter, 2007). Over more recent years, many nonprofits have formed closer ties with corporate sponsors, and they have adopted more business-like procedures. Drawing on a survey of nonprofit activity around the world, SustainAbility (2003: 51) believes that the nonprofit sector is gearing towards more market-based solutions, mechanisms and dynamics. On the other hand, businesses in general have assumed social responsibilities in a broad array of their activities, as well as responsibilities for the natural environment and natural resource consumption. We have also seen the rise of an organization type in an apparently new hybrid space, which combines social and economic goals. The literature most commonly identifies this as a 'social enterprise'. Alter (2007) sees the pioneer social entrepreneurs as: John Durand, who began the first 'social firm' with disabled employees in 1964; Mimi Silbert, who established Delancy Street social businesses for recovering addicts in the 1970s; and Mohammad Yunus, who popularized microfinance with the Grameen Bank in 1976.

There is no clear, consensual definition of 'social enterprise' in the literature – much less so for eco-social business. As far as social enterprise is concerned, it appears that each discipline or field defines in its own image. *The Nonprofit Good Practice Guide* (2009), for example, defines a social enterprise as 'a nonprofit venture that combines the passion of a social mission with the discipline, innovation and determination commonly associated with for profit businesses'. Alter (2007) takes a more market-oriented definition: 'A social enterprise is any business venture created for a social purpose – mitigating/reducing a social problem or a market failure – and to generate social value while operating with the financial discipline, innovation and determination of a private sector business'. The literature is virtually silent on developing-country-based and -financed social enterprises (Chikandi, 2010), apart from micro-finance, which is used as an exemplar of social entrepreneurship.

Our contribution enlarges these consolidated research fields, offering a theoretical framework that tries to put in a more holistic vision into the topic of eco-social business in developing countries. The framework focuses on the main aspects that should be considered for supporting the sustainable development of emergent countries. The literature recognizes several variables that influence the effectiveness of a sustainable use of resources in a developing country, but we did not find a consolidated definition of social business. There also is no clear definition of social entrepreneurship. Although entrepreneurship has long been recognized as a fundamental institution of economic growth, it has only recently begun to receive attention by development economics, which has historically favored top-down, planning-oriented strategies to poverty alleviation (McMullen, 2011). But if entrepreneurs are an important factor for developing countries, the sustainability issue requires a clear analysis of why responsible investment is made. There are three main approaches that deal with responsible investments in emerging nations: the 'bottom or base of the pyramid' approach, the 'social business' approach and the 'public purpose capitalism' approach. Even though all the approaches, to some extent, consider the 'eco-business' issue – that is, responsible care and sustainable use of local natural resources within a strategy or a set of strategies – it is thought that the sustainability perspective warrants a more encompassing view. So we offer an extended association among these approaches and some other 'ingredients' to the issue.

Based on the above, we develop a model that could lead to explaining and expounding eco-social business formation in the developing world. We will show the role of entrepreneurs and of what is known as the 'factor-four strategy' for developing eco-social businesses. Connecting this to emerging markets, the main aim is to find how this can progress in an environment with limited economic resources, scarce employment opportunities, abundance of unskilled labor, low levels of technological know-how and insufficient governmental capabilities. The answer would lie in a combined effort of integrating strategies developed by all the actors of an emergent country.

A BRIEF LITERATURE REVIEW

Eco-Social Business and the Role of Eco-Social Entrepreneurs

As for eco-social business, after the seminal publication of Bennett (1991) on the concept of 'ecopreneurship', the literature exploring the intersection of entrepreneurship with environmentally and socially responsible behavior has been relatively scarce (Randjelovic et al., 2003; Cohen, 2006). Researchers have refocused on the creation of enterprises with an environmental (and more latterly social) mission (Dickson et al., 2007; Ivanko, 2008; Schaper, 2005), but exploration of the relationship between the entrepreneurship discipline and the (natural) environment is embryonic. What has been created, nonetheless, is a variety of terms such as 'ecopreneurship' (Bennett, 1991; Schaper, 2005), 'eco-entrepreneurship' (Randjelovic et al., 2003), 'environmental entrepreneurship' (Schaltegger, 2005), 'sustainable entrepreneurship' (Masurel, 2007), 'enviropreneurship' (Menon and Menon, 1997) and 'green entrepreneurship' (Berle, 1991). These terms all describe some aspect of 'entrepreneurship through an environmental lens' (Schaltegger, 2005). The majority of publications on ecopreneurship are compiled in a reader by Schaper (2005) who defines the distinguishing characteristics of ecopreneurship as: 'entrepreneurial in some shape or form; commercial activities that have a net positive impact on the environment and move towards a sustainable future; and an intentionality where the ecopreneurs personal belief system sees environmental protection and a more sustainable future as important goals in their own right'. The incorporation of sustainability also suggests an implicit commitment to a social dimension. The inclusion of intentionality is also important as it precludes 'accidental ecopreneurs' where the environmental outcomes are accidental by-products of other business activities (Schaper, 2005).

Many of the 'ecopreneurial' theorists include social dimensions in their discussions, considering entrepreneurship as a vehicle for social and environmental change (Anderson, 1998). Another dimension covered within the entrepreneurship literature is the characteristics of the entrepreneurial individual: the classic entrepreneur who sets up a small business that grows into a successful company (Schaper, 2005) or the corporate intrapreneurs setting up new initiatives within existing organizations (Pinchot,

1985). Organizations themselves can also be considered to demonstrate entrepreneurial qualities in the way they innovate and develop new solutions (Schaper, 2005).

The focus of all the above is mainly on the United Kingdom (UK), Canada and the United States (US) (Holt, 2011), even though there exists an expanding multitude of local social enterprise hybrid organizations in the developing world which also serve ecological needs (Borzaga et al., 2008). But currently there is no comprehensive discussion involving eco-entrepreneurship that takes place in developing countries (Holt and Littlewood, 2011). Some acknowledgment to social and eco-social enterprise in the developing-world context is made in the debates involving philanthrocapitalism (Edwards, 2008). However, philanthrocapitalism is but one specific application of social enterprise in the developing world, and most importantly, it does not address or give voice to the situation of local social enterprises run and financed by local people (Hackett, 2009).

There may be one principal reason for this inattention: eco-social business in most developing countries is often barred from the outside world through complex market failures: lack of transport, infrastructure, monetary systems, legal regulations and communication channels, for example, pose formidable barriers to 'proper' functioning of a Western conception of 'the market' (Dorward et al., 2005). One result of this is high transaction costs that make production and selling of indigenous goods and services expensive, or unavailable, especially from remote rural communities. Simultaneously, there is lack of telecommunication channels resulting in asymmetric information, which also makes it difficult for farmers to get their products to the market, or get a good price. We see here what Stuart L. Hart (2007) has called the 'collision of the monetary economy, the traditional economy and nature's economy': the monetary economies of the developed countries would just place a burden on the other two economies. But, as Hart continues, he illustrates that today's digital revolution is changing the pattern and that business opportunities arise from knowledge being brought to the world of the poor which will empower them to make use of their natural and human resources (Hart, 2007: 40–41). If businesses (of the developed world) wish to expand beyond their terrain, they must focus attention on the needs of the isolated and disconnected. In this context, Hart and Milstein (2003) report on Hewlett-Packard's 'World e-Inclusion' initiative which created a research and development (R&D) laboratory in rural India with the express purpose of coming to understand the particular needs of the rural poor, and the report states that local companies such as N-Logue and Tarahaat have also developed information technology and business models focused on this enormous potential market.

It is not, however, just about spreading knowledge of technology. What we also find in developing countries is that they have broadly different local power structures than wealthier states, especially on the local level. Here, profit-generating activities are deeply embedded in the traditional elite–client power structures of their communities: 'How income was to be generated, by which types of activities, and in which niches of the market, were strongly influenced by the economic interests of the local elite, who also controlled the local government' (Makita, 2009: 53). The existence of such power structures makes a considerable difference to the development and execution of eco-social business opportunities. For example, in community development programs, benefits are often captured by local elites instead of reaching their intended (poorer) targets (Kennedy, 2011). Furthermore, local power structures often affect the mission of development agencies and of foreign investors who would seek social change activities and responsible care of natural resources. The view of the elites, if they are not corrupt (which unfortunately is the case often enough), is more focused on quickly presentable outcomes of poverty alleviation activities. Thus, if ever spending reaches rural areas, it often focuses on creating non-farm jobs as these produce wages and hence bolster labor statistics (Mutagwaba, 2009). But what really would be needed is to reach out to subsistence farmers in remote areas and to equip them with finance, training, access to technology, housing, education and healthcare. Fortunately, this mission has been taken up, for example in sub-Saharan Africa, by a number of foreign investors who have taken care of eco-social needs together with their agricultural investment. Examples are Syngenta of Switzerland, Malaysia-based Sime-Darby Investment Group and Chinese agro-businesses. Details will be given in the typology section below.

Eco-Social Business: The Sustainability Issues and Responsible Investment

From all the angles that were explored above, it emerges that sustainability issues are getting more and more important in the context of developing the economies in the world of emerging nations. This requires strategies that not only solve social and economic problems, but also care for the maintenance of resources and bring about resource consumption that benefits the local communities and their environment as well as the revenue interest of their respective national governments. Achievements can be reached by both outside investment and indigenous endeavors, that is, by entrepreneurs who have roots in the respective environment. Outside investors would be those who integrate environmental, societal and governance issues into their decision-making, and quite a few of these 'responsible investors' operate in many financial markets worldwide, through either ethical investment funds or private equity (Louche and Lydenberg, 2011).

Responsible investment emerged in the 1970s when societal and environmental activists in the US and elsewhere combined the divestment techniques of faith-based organizations that refused to profit from businesses they regarded as unethical (primarily tobacco, alcohol and gambling) with the tactics of community and consumer activists such as Saul Alinsky and Ralph Nader (Sparkes, 2002). As of today, responsible investment funds offer opportunities for a broad variety of business lines. However, their focus is more on societal and less on environmental concerns, let alone concerns for nature-related issues in developing countries (UK Social Investment Forum, 2007). This seems understandable because not even an ethical investment fund will aspire to successfully deal with, for example, the depletion of rainforests or with the pollution of water from oil spills in Nigeria. Situations like these call for an indigenous or at least an on-site solution through location-specific investment, which, if foreign, might involve a specified investor, private equity, non-governmental organizations (NGOs), charitable funds and so on.

There is another obstacle to dealing with environmental issues in the developing world: the circumstances in developing countries often contradict the idea that we can consume some of our natural capital (in the form of environmental degradation, for example) as long as we offset this loss by increasing our stock of man-made capital, making use of the technological advances. This way of thinking presupposes that technological advances can be transferred from where they originate (in the developed world) to where they are needed to offset the loss of natural capital (in the developing world). In the dominion of sustainability theory, the replacement of natural capital by man-made capital has been termed 'weak sustainability', as opposed to 'strong sustainability' which requires that the resource structure must remain unchanged (Pearce and Atkinson, 1993). The advocates of the 'strong sustainability criterion' see nature as an indivisible heritage and they reject what they call 'commodification' of the environment; in their view, the market functions as a collective action against sustainability, and extraction or production of resources adapts nature to human technology and methods, while it should be the other way round (Scherhorn, 2004). But while there may be reasonable arguments as to whether substitution of man-made capital and natural capital is moral or not, it is commonly agreed that natural capital and man-made capital should be managed at optimal levels over a longer time scale (Daly, 1990). Hart (2007) gives an account of modern agro-forestry practices in developing countries that realize high returns while minimizing environmental impact (Hart, 2007: 48–9). This may be used as an analogy for improvements in the mining of minerals which would help to relieve pressure on this type of natural capital stock.

Viewed from another perspective, environmental issues in the developing world require a paradigm of sustainability which regards collective action to be at par with entrepreneurial action: cooperative actions for sustainability operating on both the government level and the level of single firms will bring about conditions that enable producers and consumers to contribute to sustainable development rather than interfering with it. Such actions will have an optimum outcome if they do not harm the so-called 'essential' (Dobson, 2000) or 'critical' natural capital (Neumayer, 1999) – air, biodiversity, climate, soil, and water – and thus maintain the earth's ecosystem services (Dyllick and Hockerts, 2002: 133). This would be achieved by obtaining energy from renewable sources such as solar heat or light, wind or water power, or solar-generated hydrogen, and by recycling and reuse of non-renewable materials including fossil fuels, which means that waste and emissions are either avoided or upcycled. Upcycling denotes a kind of recycling that produces materials of at least equal quality (Pauli, 2000; Braungart and McDonough, 2002), while downcycling involves converting materials and products into new materials of lesser quality. In developing countries, where new raw materials are often expensive, upcycling is commonly practiced, largely due to impoverished conditions. But in both developed and developing countries there is enormous potential in making proper choices for this type of technology, and the outcomes are often innovative processes and products (Wang, 2011).

Recycling, reuse and upcycling would be the logic of the 'factor-four strategy' (Von Weizsäcker et al., 1997): halving resource use and doubling wealth, that is, decreasing the devastation of nature, and at the same time increasing private welfare. That strategy will perform at its best in what has been called 'metabolic settings', that is, viewing the production–consumption process as a circulation driven by renewable energy (Scherhorn, 2004). Thus, the improvement of the production and recycling processes will promote a feeling of caring for nature and may push back the tendency of individuals or enterprises to pass on their costs to the environment. From there, mankind would also get a better understanding of how the quality of life can be improved by taking care of nature.

But how can a 'factor-four strategy' be put into practice in an environment with limited economic resources, scarce employment opportunities, abundance of unskilled labor, low levels of technological know-how and insufficient governmental capabilities? The answer would lie in a combined effort of integrating strategies from the 'bottom or base of the pyramid' and 'public purpose capitalism' approaches with cooperative actions for sustainability. This will be set forth in the next section.

FROM BOTTOM/BASE OF THE PYRAMID STRATEGIES TO PUBLIC PURPOSE CAPITALISM: COOPERATIVE ACTIONS FOR SUSTAINABILITY

The approach known as 'bottom or base of the pyramid' (BOP) was coined by C.K. Prahalad and Stuart L. Hart (2002). The main observation of the approach is that 4 billion people remain outside of the global market system; hence, new ways of doing business will have to be introduced into this market. However, tapping into these overlooked markets will mean that investors reconfigure their business assumptions and strategies. They need to identify a new segment of customers, and even though these segments provide only low margins, unit sales can be very high, and from there, the product/service can be an important source of income. The argument is that selling to the poor will not only benefit them as consumers, but also enable them to reap the benefits of respect, choice and self-esteem and provide opportunities to escape from poverty (Prahalad, 2009). The BOP approach is both top-down, with main attention on poor people's capacity to consume; and bottom-up, advocating that local knowledge should be increased, additional capabilities should be developed and all activities should be socially embedded (Ackermann, 2010). Embeddedness would have to go as far as to include the activities in the informal sector since it has been estimated that well over half of the total economic activity in the developing world takes place in informal sector relationships, which are grounded primarily in social instead of legal contracts (de Soto, 2000). But while one may argue that a defining characteristic of those at the bottom of the pyramid is that they are not integrated into the formal global economy (Hammond et al., 2007), it seems too negative to believe that all business activities conducted by the poor are likely to remain in the informal economy for the foreseeable future. There are definitely chances to connect them to global markets.

Connecting to global markets and still staying well embedded in the social – and natural – environment would be the way that leads to social and eco-social business. But it also means that foreign businesses no longer sell just to the poor, but also enter into co-ventures with local businesses (Simanis and Hart, 2008). This co-venturing will encourage the build-up of new capabilities which would then create their own markets (such as services, for example in the tourist industry or in communications, that were hitherto inconceivable). Involvements of this type should also reach the lowest segment of the pyramid and allow people to benefit from trade-offs or consume more or raise their income (Karnani, 2005). What we have here is a wider aspect of what has been called the 'linkage phenomenon': the original notion of linkage effects from (foreign) investment comes from Hirschman who stated in 1958 that spillovers between different industries played an important role in development (Hirschman, 1958). Since then, a broad discussion has evolved on the supplementary effects of investment, especially with a view to developing countries. The effects are much broader than what is occurring just between industries, because the effects of new business activities in an environment that has had no business at all (or just 'informal' business transactions) will provoke changes in thinking, even in values, and in social order (Buckley, 2009).

A practical example of how the linkage phenomenon extends to investment spillovers towards social and ecological issues has been displayed through 'public purpose capitalism', a term that was coined by Andrew Young (Sehgal, 2010) to denominate socially active public–private partnerships where the private sector would invest in public-purpose facilities, buildings and businesses. The concept had a tremendous success in Atlanta, where Andrew Young was the town's mayor from 1981 through 1989, and it was then transferred to Africa where the Southern Africa Enterprise Development Fund (SAEDF) was founded in order to enhance social and infrastructure projects. The most recent application of the concept is in Haiti, where a similar Enterprise Development Fund is being created which it is hoped will produce a sustainable economy after the earthquake disaster, and not just a relief economy aided through grants from abroad (Young, 2010).

The United Nations Conference on Trade and Development (UNCTAD, 2007) states that the presumed benefits of foreign investment spillovers are primarily in soft technology, that is, organizational, managerial, information, processing and other skills and knowledge, as well as more efficient inputs into primary and manufacturing industries and linkages to global markets. Again, effects on the environment have not been explicitly recorded. But when foreign investment engages in sectors that are closest to domestic resources, they bear a high potential for influence not only on the local economy but also on the social order and the natural environment (Rugraff et al., 2009).

The predominant effect that is conceptualized via the level of foreign engagement is poverty reduction (Rugraff et al., 2009; Jeppesen and Wad, 2006). But the impacts go further: both the BOP approaches as well as the social business approach, while aiming primarily to reduce poverty, deal with living conditions and they also affect the local natural environment of these communities. Here we can see this wider type of 'linkage', as it goes well beyond just economic consequences. Quite a few of the social businesses promoted by microfinance also have ecological consequences (Yunus, 2010), and with regard to BOP, Prahalad (2009) has exhibited three examples where ecological considerations are an explicit part of an investor's mission. One is EID Parry, the Indian sugar producer which supports farmers not only with access to fertilizers and tools, education and crop disease diagnosis, but also with technology for protecting soil and for minimizing pollution. The second case is CEMEX, the world's third-largest cement producer, which combines social objectives (homes for the poor) with environmental objectives (less flooding and mud, less dust, and so on) in its Mexican 'Mejora Tu Calle' projects. The third one is Tecnosol, which provides clean energy alternatives for the lighting and refrigeration needs of rural Nicaraguan households, schools and hospitals that have no access to the main electricity grid. The spillovers from these and other investments not only raise the level of skills and provide sources of income in adjacent business lines, but they also improve general living conditions and awareness of environmental protection, by which they eventually generate new attitudes and even new value systems.

This perspective of a wider 'linkage' effect may now lead us to a theoretical approach that explains the underpinning of eco-social business creation in developing countries.

A THEORETICAL FOUNDATION FOR CREATING ECO-SOCIAL BUSINESS IN DEVELOPING COUNTRIES

The scheme that has been identified as 'social business' promotes the idea of doing business in order to address a social problem, and not to maximize profit. This goes along with the statement that 'the business of business is, increasingly, the creation of social value together with economic value' (Austin et al., 2006). Social actions are no longer considered to be the exclusive realm of not-for-profit organizations and governments; and conversely, many not-for-profit organizations are striving to gain new managerial skills that may allow them to improve their social development performance. And while Yunus's initiatives and further endeavors relate to small business start-ups in the developing world, there are now novel income-generating activities which arise from what conventional firms and not-for-profit organizations view as a new part of their mission. With regard to the not-for-profit institutions, the entrepreneurial nature of these initiatives aims at achieving management efficiency and profit generation or at least financial self-sufficiency (Alter, 2007; Borzaga et al., 2008; Noya, 2009). The initiatives may be considered as additional funding mechanisms or as program mechanisms in support of the organization's mission (Masi, 2009). From the other angle, the for-profit business has adopted a new sense of social and environmental responsibility. The 'new' in this is the combination of both environmental responsibility, which has been present in, for example, the chemical industry for quite some time with the Responsible Care© initiative having been launched long before the term 'sustainable development' was adopted by the UN; and social care has been on almost any employer's agenda right from when a business is initiated.

From the above, we derive the number one theoretical ingredient for eco-social business, and this would be the 'mission' of the social entrepreneur (1); to which we would have to add, as the other ingredients, the theory of the factor-four strategy of sustainable development (2); an expanded concept of linkages (3); and encouragement from pertinent policy agendas (4).

The Mission of the Social Entrepreneur

Three kinds of eco-social entrepreneurs are needed, based on their roles and working environments: policy, program and business entrepreneurs. At the macro level, social entrepreneurs could help formulate and implement policy; at the business level, they could use their business skills to address social issues; and at the community level, they could help solve specific local problems (Babu and Pinstrup-Andersen, 2007).

Firstly, policy entrepreneurship. Social entrepreneurs well versed in policy processes expand successful local programs into large-scale national programs with a wider poverty impact. Bringing about significant changes in policy at the national or global level, however, requires change agents at the highest levels of decision-making. At the global level, policy entrepreneurs could influence policy-making by multilateral aid agencies; at the national level, they could guide national systems toward specific strategies; and at the local level, although they should limit their influence to mere counseling, they could help create a policy environment that enables other types of social entrepreneurs to be effective.

Secondly, program entrepreneurship. Program entrepreneurs are instrumental in designing and implementing innovative programs to reach the threefold objectives of alleviating poverty, developing economic opportunities and conserving the environment. They help to implement programs that are funded by development partners, national governments, and NGOs in a way that addresses local problems with global ideas. Their main allies would be youth and youth leaders, as these are increasingly seen as partners in development. On the one hand, many youth are engaged in community affairs, have a high level of commitment, and are well connected through information and communications technologies. On the other hand, involving the growing number of educated but unemployed youth that exists in many countries will decrease the risk of social instability and armed conflict. Given appropriate skills, mentoring, recognition and support, these individuals could become effective social entrepreneurs, and their engagement and collective action could be transformed from negative to positive action. This holds true particularly in transition economies, where young leaders are needed who 'unlearn' the beliefs of their forerunners (Kennedy, 2011).

Thirdly, business entrepreneurship. Social business entrepreneurs encompass business leaders who are successful in their own field and employ their business acumen in solving social problems; for example, a commercially successful physician who organizes fellow doctors to provide health services to the rural poor at no cost or minimal cost. As with the BOP strategies, another group of social business entrepreneurs would be a subset of poor people, who – although they all fall below the poverty line – may still have some level of income, resource ownership, social capital and entrepreneurial abilities. This group is increasingly engaging in self-reliant efforts, as those poor people and poor societies borrow ideas and institutions from the West when it suits them to do so (Easterly, 2008). This coincides with the new attention entrepreneurship has received from development economics, where it is argued that the societies of the bottom billion may best be rescued and transformed endogenously through bottom-up, market-based strategies (McMullen, 2011).

Those three types of social entrepreneurs roughly correspond with what McMullen (2011) has called business entrepreneurship, social entrepreneurship and institutional entrepreneurship in his new theory of development entrepreneurship. Similarly, Zahra et al. (2009), asking about the motivation of the social entrepreneur, depict the 'social bricoleur', who uses the knowledge of a local community to solve a communal issue; the 'social constructionist', who addresses social needs that cannot be fulfilled by government; and the 'social engineer', who focuses on issues beyond the national level and aims to redesign or replace globally existing social structures. Analyzing these phenomena would dig much deeper than we wish to chart here since our focus is on the sustainability aspects. It may suffice to point out that all those types of entrepreneurial activity will help to develop productive capabilities and contribute to improving the quality of life in the social environment where the activities take place. This definitely requires management efficiency, but one other condition would be the availability of technology and of appropriate skills.

Applying the Factor-Four Concept

Technology plays a foremost role in the factor-four concept of sustainable development that was mentioned above, as it moves away from labor productivity and towards resource productivity. By using best available technology, advanced engineering and improved production methods, fewer resources are required to produce more products and services. The aim is for society to last twice as long or enjoy twice as much, whilst using half the resources and placing half the pressure on the environment.

Factor-four is used in decision making, production and product-oriented environmental protection. Fundamentally, factor-four is an economic idea. Reducing resource use by a factor of two and output by again a factor of two is not a fixed target. In their book, which was produced as a report to the Club of Rome, von Weizsäcker et al. (1997) point out that, in some cases, well-employed technologies and processes can increase resource efficiency much more. In other cases, technical limitations are given. So the number 'four' would be some kind of an average goal, but this does not mean that is even a stretch goal. The objective is to expand the lifespan of resources as much as possible so that more wealth can be extracted from the resources we currently use. Since 'more wealth' would be what is primordially needed in transition economies, applying this concept to projects that take place in these countries should be a primordial target. The ways to get there and their effects can be demonstrated by a glance at the main imperatives for action in the two problem areas of 'source-side' and 'sink-side' issues of sustainability (Sanya, 2007):

Source-Side Problems
Depletion of renewable and non-renewable resources:

- Make efficient use of environmental inputs so as to minimize resource consumption in production processes
- Make sustainable use of renewable resources to allow them time to regenerate naturally
- Substitute non-renewable resources with renewable resources so as to preserve the non- renewables

Loss of environmental quality:

- Preserve and regenerate environmental quality and biodiversity

Sink-Side Problems
Increased pollution with negative consequences such as global warming

- Use clean production methods that minimize pollution

Accumulation of human-generated solid waste

- Reduce, reuse and recycle waste. Adopt closed-loop systems whereby the would-be-waste becomes a useful production input
- Make products, which, at the end of their useful life, will turn into waste that will be easily degradable and non-toxic to the environment.

Both Source- and Sink-Side

- Reduce overall demand for environmental resources by controlling population growth and limiting materialistic consumption.

There are numerous cases, for example in Africa, in which these imperatives for action have yet to be set, from the neglect of horrendous oil spills in Nigeria and the tremendous gap between the expectations created by oil riches and the reality (Gary and Karl, 2003), to the massive discharge of solid, liquid and gaseous effluents directly into the lagoons at the Gulf of Guinea (Oshisanya and Oshisanya, 2009), to the forceful alienation of communities that should partake in the exploitation of the productive resources of their lands (Barney Pityana, 2003). From there, both the academic and business worlds have recently been paying growing attention to the field of green and clean technology implementation in developing countries. The main investments were dedicated to the renewable energy sector from the late 1990s and to clean water (Piebalgs, 2007; Leitner et al., 2010; Wüstenhagen et al., 2008). It is widely acknowledged that the diffusion of clean technology innovations in the developing world not only requires changes in user practices, policy and cultural discourses, as well as government institutions, but also new forms of entrepreneurship (Aldrich and Ruef, 2006). But apart from this strand of 'high-tech' we also find increasing evidence for advances produced in traditional industries such as the 'African Cashew Initiative' (ACI, 2010). This is a multi-stakeholder project which demonstrates that applying proper technologies and skills is just one instrument to improve the situation: it will only produce its direct effects and the desired effects of spillover and linkage if all communities that are affected become actively involved.

Expanded Linkages and Community-Based Participation

If investment spillovers are to raise the level of skills, provide sources of income in adjacent businesses, improve general living conditions and awareness for environmental protection, and even generate new attitudes or enhance new value systems, some kind of 'receptacle' is needed to respond to these effects. This absorption capacity is highly dependent on the level of capabilities, and a certain threshold level is needed for this

(Perrons, 2004). The level can be raised considerably if the respective communities become actively involved, and it will often be necessary that social structures be changed to reach this level. Thus, community-based participation and organization – as well as people's right to play an active role in their own process of sustainable human development – is one of the foremost requirements for linkages that reach further. Communities often do not participate when programs, projects and activities do not meet their desires, needs and demands. Even more so, outsiders who come into communities without adopting participatory approaches will end up imposing projects that do not address the needs of those communities or relate to their dreams. The results will be in low levels of mobilization and participation. But when a proper approach is adopted, cross-national and international cooperation and capacity building combined with adequate resource management practice can reach out to generate employment and income in communities that live quite distant from more developed ones.

A good example is the success of three tribes who live in remote areas in Brazil, where conservation zones have been set up in Amapá and in the Amazonas states by the national environmental protection agency (Inoue and Do Prado Lima, 2007). The people who live there tend to be poor and have almost no access to education and healthcare and very little access to energy, transport and communication due to deficiencies in the infrastructure. They rely largely on harvesting plants, fishing, hunting and small-scale farming, which are usually intended for subsistence or for making sales in local markets and only provide enough resources for basic survival. Job creation and an increase of income that reaches beyond subsistence economies can only be achieved by understanding economic dynamics beyond the local level and by seeking out secure and sustainable means of gaining access to market economies. The indigenous tribes in question received appropriate assistance from ecologists and social entrepreneurs and have learned to sell part of their production in regional or national markets. In another case, rural communities in Brazil started to cultivate an oil seed which had not been used for anything but firewood, with a fieldwork advice from a Bavarian agricultural cooperative. This plant was harvested and the seeds were converted into a product that could be shipped overseas. Similar examples from Africa will be presented in the 'Typologies' section below.

Encouragement from Policy Frameworks

Discussions on sustainable use of natural resources fit into the context of the global biodiversity regime and, more specifically, debates on how to balance conservation, development and improvement of the quality of life of local populations. Through the signing of the Climate and Biodiversity Convention (CBD) at the United Nations Conference on the Environment and Development, held in Rio de Janeiro in 1992, a new environmentalism was formed which could be founded on international regulation (McCormick, 1992). The Convention attempted to reconcile preservation, conservation and economic development, and to accommodate the interests of governments, NGOs and grass-roots movements from countries in the North and South. The CBD also built bridges between the natural and social dimensions of biodiversity-related issues, with the objectives being: (a) to conserve biodiversity through the sustainable use of its components; and (b) to achieve the fair and equitable sharing of the benefits derived from the use of genetic resources. The Convention and its subsequent regulatory updates make a distinction between conservation and sustainable use of biodiversity. This was done in response to developing countries wishing to clarify that conservation and sustainable use of biological resources are different but complementary objectives (Alencar, 1995).

Before the CBD, an earlier effort to commonly address social, economic and ecological issues was launched by the International Union for Conservation of Nature (IUCN), the United Nations Environment Programme (UNEP) and the World Wildlife Fund (WWF) in 1980 under the label of 'World Conservation Strategy' (WCS), and it was acknowledged that it is not possible to conserve nature without considering the needs of human populations: 'People whose very survival is precarious and whose prospects of even temporary prosperity are bleak cannot be expected to respond sympathetically to calls to subordinate their acute short term needs to the possibility of long term returns. Sustainable development must therefore include measures to meet short- term economic needs' (IUCN, UNEP and WWF, 1980: Item 11).

Other approaches were labeled as 'community-based conservation' (CBC), 'integrated conservation and development projects' (ICDP), 'community-based wildlife management' (CWM) and 'community-based natural resources management' (CBNRM). These participatory approaches have sought to seek local definitions for environmental problems and solutions, and they have promoted the role of traditional knowledge and of resource management for local needs (Jeanrenaud, 2002).

All these frameworks and agendas as well as the regulatory systems that they have spurred on the supranational and national levels were very helpful with regard to how eco-social business was accepted in the economic environment. Even though a formal regime of support for eco-social business was not delivered in any of the countries that were signatories to the 1992 United Nations Conference on Sustainable Development in Rio de Janeiro, the resolutions of this conference, contained in what came to be known as Agenda 21, have at least fostered eco-social entrepreneurship (Hemmati et al., 2002).

APPLICATION OF THE MODEL TO EMPIRICAL CASES OF ECO-SOCIAL BUSINESS (A TYPOLOGY)

With the four 'ingredients' of eco-social business in developing countries (social entrepreneurship, sustainable development, an expanded concept of linkages and encouragement from policy agendas) as per our model, we can now apply the model to some empirical cases. In order to facilitate the analysis we have clustered our cases in a typology. By delineating from incidence that can be studied in the 'real world', foreign or at least non-local intervention may be viewed as one criterion for classification. Thus, five types of eco-social business in developing countries can be categorized: one type is fully launched through foreign investment by business firms; a second type is prompted by foreign not-for-profit businesses; a third type is science-based, using technology brought to a host country from abroad by some sort of research and development agreement; another would relate to eco-tourism which often involves cross-border business; and a fifth type is fully local and not created, or at least not directly generated through foreign intervention.

Firstly, eco-social businesses that are prompted through foreign investment. There are numerous cases of transnational companies (TNCs) which have set up specific programs to achieve eco-social objectives together with a long-term financial return on their investment. They train farming people in rural areas of Africa and provide healthcare and housing for crop pickers, such as the Malaysian palm oil producer Sime-Darby (Brown, 2011) and the Swiss agrochemicals producer Syngenta (Stone, 2009); or they actively seek to combine business progress and social progress in all their agro-industrial ventures, such as Danone (Das Gupta, 2011; Ghalib and Hossain, 2008). And contrary to what has often been found about China's investment in Africa, Chinese investors have carried out a multitude of agricultural projects in Rwanda and Sierra Leone that went down to the village level. Hundreds of rural households were involved in projects on water control, fishery, crops, livestock and veterinary services, and thus started to 'unlearn' old habits and acquire new ones (Spring, 2009). On the issue of technology, there are quite a few efforts to enhance agricultural productivity in a prudent way and step by step, which will create substantial employment opportunities to the teeming population in the rural sector and up the value chain, and will terminally reduce the poverty level in Africa (Obot and Obot, 2009). Another example is low-carbon foreign direct investment (FDI) which flows into three key low-carbon business areas (renewables, recycling and low-carbon technology manufacturing). For developing countries, low-carbon FDI can facilitate the expansion and upgrading of their productive capacities and export competitiveness and help their transition to a low-carbon economy, with both short-term and long-term economic and social advantages (UNCTAD, 2010).

Secondly, eco-social businesses launched through foreign and non-foreign not-for-profit institutions. While being not-for-profit organizations by nature, well-structured NGOs will have the knowledge, experience and skills needed for pursuing management efficiency. They would therefore provide the technical, social and economic dimensions needed in governing and managing social enterprises. This may be also achieved by means of close alliances between those NGOs and private businesses that would go beyond advice, trade and financial agreements (Ruli and Hoxha, 2001; Alter, 2002) and would lead both organizations to be actively engaged in governing the enterprise. There is a set of cases from Italy and Peru, where the Italian NGO Cesvi (an acronym which means 'cooperation and development') has been operating since 1989 (Masi, 2009). Cesvi's primary aim had been to foster disadvantaged people's social integration through work, for example for victims of infantile prostitution in Lima. This has evolved into a business concerned with the production of bakery and pastry products by using natural ingredients. In a second initiative, teenagers and young women that had been victims of sexual aggression received job training as dressmakers in Peru's capital Lima and a co-operative was built for cloth making, particularly work uniforms, from natural fibers. Another case is reported from the Swiss business firm Greentecno which is dedicated to the development of rural communities in Africa, providing hybrid wind–solar generators and devices producing drinking water from atmospheric humidity. The company has joined forces with various NGOs and local communities operating on 'ethical projects' especially in South Africa, and it uses microfinance mechanisms to establish local independent power-producers (Molteni and Masi, 2009).

Thirdly, science-based eco-social businesses adopting outcomes from foreign R&D. Collaboration between scientific institutions has become a vehicle to promote businesses in developing countries including eco-social entrepreneurship. One example is the cooperation between the US National Academies and the Nigerian Academy of Science (Committee on Creation of Science-Based Industries in Developing Countries, National Research Council, 2007). In Nigeria, as in many other African countries, with about two-thirds of the population lacking basic services such as electric power, safe water and access to effective medicines for infectious diseases, the government is unable to supply solutions for these issues in a timely and enduring manner. While in developing countries, lack of safe water and lack of home or small business electric lighting have generated entrepreneurial solutions through readily accessible technologies, in other countries, of which Nigeria is an example, private companies wishing to extend basic services to the underserved have generally not been viewed as an instrument of government policy. But the country has proven that adopting outcomes from foreign R&D, when properly channeled, can encourage private companies to provide these services in a sustained format. In the Nigerian case, three technologies were explored by the two science academies: solar photovoltaic, water purification and effective malaria therapy, and associated business models were selected to demonstrate how the government-sponsored participation of the affected communities would create private sector enterprises that provide the services (Committee on Creation of Science-Based Industries in Developing Countries, National Research Council, 2007).

Fourthly, eco-tourism businesses through cross-border investment. Eco-tourism is estimated by a 2010 report to have revenues of $28.8 billion a year in developing countries alone, and the net present value of eco-tourism-controlled land as represented by the producer surplus (profits plus fixed costs of eco-tourism lodges) is US$1158 per hectare, which is higher than all currently practiced alternatives, including unsustainable logging, ranching and agriculture (Kirkby et al., 2010). Also, it creates jobs in areas that do not offer many other employment opportunities, and the type of land use connected with eco-tourism definitely protects nature. The report that was just mentioned draws on a Peruvian business with now 37 eco-tourism operations in the southwest Amazon eco-region and the tropical Andes (Callard, 2011). Eco-tourism in Africa has a longer history (Binns and Nel, 2002; Mowforth and Munt, 1998). One well-known case is the the Canadian Kellerman Foundation's long involvement with the protection of the gorillas in the Gorilla Park that is located in Rwanda and borders the countries of Uganda, the Democratic Republic of the Congo and Rwanda (Butynski and Kalina, 1998). It has launched a new project to examine the worldwide best practices in eco-tourism and micro-enterprise development so as to provide concrete recommendations for the development of eco-tourism and micro-enterprises around and within the park (Ponnekanti, 2011). An example from Asia is eco-tourism in Sri Lanka, where the Japanese initiative 'Action for Peace, Capability and Sustainability' (APCAS) has developed an agenda to attract visitors and tourist specialists from Japan.[1]

Fifthly, fully indigenous eco-social businesses. With this, we refer to a type of entrepreneurial activity that may stem from the absence of outward communication in landlocked areas, where a majority of the world's poorer population lives. What we find here is a strong environmental orientation deeply rooted in nature and nature's principles. This orientation, having been applied to rudimentary entrepreneurship from the outset (Campbell, 2006, has named it 'grounded' entrepreneurial activity) is still inherent in more advanced forms of business which have developed as those areas have reached a certain degree of professionalization in ecological matters. The results are a mutually beneficial interaction between and among individuals and between people and nature. There are examples from Mexico of highly successful endeavors in permaculture, that is, the design of human settlements and agricultural systems that are modeled on the relationships found in natural ecologies (Diver, 2002). They move farming and other agricultural industry towards more localized, more energy-efficient and less wasteful production and distribution of food. Also in Mexico, influential attempts have been made to rediscover local building materials (*viviendas ambientales*) in the context of low-energy building (Elsen, 2011; Hermoso de Mendoza, 2011). One may regard these developments as 'fertile chaos', as they just about serve to acknowledge a need for change (Schieffer and Lessem, 2009), but in many ways they seem to be well structured enough to build resilience and to pave the way for a renaissance of entrepreneurial focus (Hopkins, 2008). Ecologically grounded architecture is a topic in Uganda also, and one may relate the Ugandan poverty situation to deficiencies in the country's architecture, where houses are only available to the urban formal sector, while the informal sector lives in the poorest dwellings. There are now attempts to model the economic production of housing according to a sustainable development approach, using the centuries-old technology of building with clay (adobe, wattle and daub) and compressed earth blocks, and thus to improve social conditions while avoiding environmental problems (Sanya, 2007). The technology builds on indigenous knowledge that, while having been preserved through generations, has not been applied consequentially. Another issue of indigenous knowledge in developing countries is that it can create and exploit new business opportunities. This knowledge may be based on centuries of observation, continually developing in response to changing social and environmental conditions. An example is the increasing market for nontimber forest products, such as *Prunus africana, Harpagophytum procumbens* (devil's claw) and *Kigelia africana* (African sausage tree). Trade in devil's claw, a traditional medicinal plant, now supports a US$100 million industry. In the beginning only a fraction of the benefits went to domestic producers, while the bulk went to processors and distributors. However, some prudent low investment in improving community skills and gaining access to relevant information is now slowly changing that pattern of benefits (Katerere and Mohamed-Katerere, 2005).

This list of eco-social business variants in developing countries was meant to distinguish the relevance of foreign intervention. However, the differentiation will always be blurred by the fact that in a globally connected world, the origins of influences cannot really be separated. Yet FDI always affects local value and idea systems. In the typology used above, a first chain of the sequence might be found in eco-social businesses that are prompted through foreign investment. The inflows of capital and knowledge into a host country will at first produce new forms of communication between the stakeholders involved in an investment project. Then technology and skills will be transferred not only to the investor's business partners, but also, by way of spillovers, to civil society as a whole. Not only will education and professionalization improve and enlarge the human capital base in the host country, but also opportunities will be forged for a new kind of interaction between the recipient of the investment, the employees and the civil society as a whole, with effects on living standards and on ethical judgments, which will also relate to ecological concerns. For example, when Chinese investors in Africa improve roads and telecommunication and educational institutions and clean wasteland before they dig oil or minerals, they lay the ground for these effects (Williams et al., 2009; Sautman and Hairong, 2009). With the entry mode and the objectives for foreign investment having changed from mere resource-seeking and efficiency to a more participative approach, as depicted in several examples from sub-Saharan Africa (Akinboade et al., 2006), subsequent investments will have to become even more densely embedded in the community (Bardy et al., 2012).

At the other end of the spectrum, fully indigenous eco-social businesses receive bottom-up promotion through participatory patterns on the local level. In many African countries, the village has been revived as the basic unit of administration in the rural areas as centrally led structural adjustment programs are now being reinstated with a new focus on the community (Bar-on and Prinsen, 1999). To gain a better understanding of this we must remember that socialism was adopted by most African governments. Socialist philosophies dominated the region from the end of the 1960s to the early 1990s, and this was understood to mean that government control of all natural resources would be for the benefit of the people. The emphasis was on development, rather than on residualism, which means that subsistence-level agriculture at the village level, even though it was (and still is) the prevailing source of income, was left out of mainstream policy-making. But African tradition has long-standing means of bottom-up decision-making and concocting common ideas. In Botswana, for example, the 'Kgotla' is the central decision-making agency of a village and serves as the village's administrative and judicial center. It is presided over by the local chief, and all adult community members are expected to attend to discuss public affairs (Silitshena, 1992). The Zulu and Xhosa as well as the Swazi use 'Indaba' or 'Indzaba' to make people get together to sort out the problems that affect them all, where everyone has a voice and where there is an attempt to find a common mind. The word, in their languages, means 'business' or 'matter'. Following this village-based approach to policy formulation, together with the importance attributed to social networks in Africa, realistic income-generating schemes are generated with outputs that include respect for nature and the community and still go beyond ensuring mere physical survival (Osei-Hwedie and Bar-on, 1999).

Another approach to classify eco-social enterprises in the developing world is being undertaken by the 'Trickle Up' research project (2011–13), a systematic evaluation of eco-social enterprises across Southern and Eastern Africa which involves the development of a directory of social and environmental enterprises across the two regions. The research project uses cluster analysis and will be used to attain a taxonomy of these types of enterprises. A first exploratory categorization was recently presented at the Conference on the Business of Social and Environmental Innovation in Cape Town in November 2011 (Holt and Littlewood, 2011).

CONCLUSION

The notion of sustainability is often seen to basically encapsulate just the growing concern for the environment and natural resources. Critics have reasonably objected that the social issues of poverty alleviation, improvement of health, of living standards and of the level of education seem to be excluded. They argue that sustainability has to be to be considered as a contested concept, a concept that is 'socially constructed' and necessarily reflects the interests of those involved. One solution to this is the 'social entrepreneur', who looks for innovative solutions to society's most pressing social problems rather than leaving societal needs to the government or business sectors. The quest of social entrepreneurs may range from small steps by solving imminent problems to major changes by persuading entire societies to take new leaps (Bornstein, 2004). Eco-social business strives to combine those two streams, and we find that the most promising way for this is 'bottom-up' endeavors. This falls in line with the argument of Yunus, who says that the trickle-down effect of reducing poverty through programs induced by governments or aid agencies is unreliable. Yunus's Grameen Bank among other projects has proved that bottom-up business models can enable the poor to lift themselves from poverty and still maintain a sustainable use of resources (Yunus, 2010).

We have built a framework for what would be the elements to explain eco-social business formation (in the developing world), and we found that the major components would have to be: (1) the clearly defined mission of the social entrepreneur; (2) a well-devised vision of technologically feasible sustainability improvements as given by the factor-four concept; (3) the prudent deployment and absorption of widely scored linkage effects; and (4) the active involvement of local communities. We have given a number of practical examples that fit into a typology built on this framework. They show that sustainable value can be created out of a wide range of varying endeavors, to the benefit of all stakeholders involved, and that technology innovation, new skills and community involvement can play a dominant role in fostering social entrepreneurship. A similar nexus is required in the theoretical field, for which several research domains must combine their efforts: international business, investment, development and entrepreneurship theory, as well as social order and ethics hypotheses, to name the most primary, are explanatory fundaments for a representation of what the 'real world' brings about in the field of eco-social business. We have connected these practical results to theory, and we think that this has pointed to a way for further research. One major stream of research could be to demonstrate that social, human and environmental benefits must achieve the same rank as financial benefits when it comes to determining return on investment, and the research should demonstrate how this is to be measured.

NOTE

1. For more details on Action for Peace, Capability and Sustainability (APCAS) visit http://www.apcas.jpn.org.

REFERENCES

Ackermann, C. (2010), 'Reducing poverty: MNCs business strategies in developing countries', Master's thesis, Copenhagen Business School.

Akinboade, O.A., F.K. Siebrits and E.W. Niedermeier-Roussot (2006), 'Foreign direct investment in South Africa', in I. Ajayi (ed.), *Foreign Direct Investment in Sub-Saharan Africa. Origins, Targets Impact and Potential*, Nairobi: African Economic Research Consortium, pp. 177–208.

Aldrich, H. and M. Ruef (2006), *Organizations Evolving*, London: Sage.

Alter, S.K. (2002), *Case Studies in Social Enterprise: Counterpart International's Experience*, Washington, DC: Counterpart International.

Anderson, A.R. (1998), 'Cultivating the Garden of Eden: environmental entrepreneuring', *Journal of Organisational Change Management*, **11**(2): 135–44.

Austin, J.E., R. Gutierrez, R.E. Ogliastri and E. Reficco (eds) (2006), *Effective Management of Social Enterprises: Lessons from Businesses and Civil Society Organizations in Iberoamerica*, Cambridge, MA: David Rockefeller Center for Latin American Studies, Harvard University.

Babu, S. and P. Pinstrup-Andersen (2007), 'Social innovation and entrepreneurship: developing capacity to reduce poverty and hunger', *2020 FOCUS BRIEF on the World's Poor and Hungry People*, Washington, DC: International Food Policy Research Institute.

Bardy, R., S. Drew and T.F. Kennedy (2012), 'Foreign investment and ethics: how to contribute to social responsibility by doing business in less-developed countries', *Journal of Business Ethics*, **106**(3): 267–82.

Barney Pityana, N. (ed.) (2003), *Report of the African Commission's Working Group on Indigenous Populations/Communities*, African Commission Document DOC/OS(XXXIV)/345, Niamey, Niger.

Bar-on, A. and G. Prinsen (1999), 'Participatory planning: counter-balancing centralisation', *Journal of Social Development in Africa*, **14**(1): 101–19.

Bennett, S.J. (1991), *Ecopreneuring: The Complete Guide to Small Business Opportunities from the Environmental Revolution*, John Wiley: New York.

Berle, G. (1991), *The Green Entrepreneur: Business Opportunities that can Save the Earth and Make you Money*, Blue Ridge Summit, PA: Liberty Hall Press.

Binns, T. and E. Nel (2002), 'Tourism as a local development strategy in South Africa', *Geographical Journal*, **168**: 235–47.

Bornstein, D. (2004), *How to Change the World: Social Entrepreneurs and the Power of New Ideas*, New York: Oxford University Press.

Borzaga, C., G. Galera and R. Nogales (eds) (2008), *Social Enterprise: A New Model for Poverty Reduction and Employment Generation – An Examination of the Concept and Practice in Europe and the Commonwealth of Independent States*, Bratislava, Slovak Republic: UNDP Regional Centre for Europe and CIS.

Braungart, M. and W. McDonough (2002), *Cradle to Cradle: Remaking the Way We Make Things*, New York: Farrar, Straus & Giroux.

Buckley, P.J. (2009), 'The impact of the global factory on economic development', *Journal of World Business*, **44**(2): 131–43.

Butynski, T. and J. Kalina (1998), 'Gorilla tourism: a critical look', in E.J. Milner-Gulland and R. Mace (eds), *Conservation of Biological Resources*, Oxford: Blackwell Science, pp. 294–388.

Campbell, K. (2006), 'Women, Mother Earth and the business of living', in C. Steyaert and D. Hjorth (eds), *Entrepreneurship as Social Change*, A Third Movements in Entrepreneurship Book, Cheltenham, UK and Northampton, MA, USA: Edward Elgar Publishing, pp. 165–87.

Cohen, B. (2006), 'Sustainable valley entrepreneurial ecosystems', *Business Strategy and the Environment*, **15**(1): 1–14.

Committee on Creation of Science-Based Industries in Developing Countries, National Research Council (2007), *Mobilizing Science-Based Enterprises for Energy, Water, and Medicines in Nigeria*, Washington, DC: NRC.

Daly, H. (1990), 'Commentary: toward some operational principles of sustainable development', *Ecological Economics*, **2**: 1–6.

De Soto, H. (2000), *The Mystery of Capital: Why Capitalism Triumphs in the West and Fails Everywhere Else*, New York: Basic Books.

Dickson, B., D. Watkins and J. Foxall (2007), *The Working Partnership: SMEs and Biodiversity*, Cambridge: Fauna and Flora International.

Diver, S. (2002), *Introduction to Permaculture: Concepts and Resource*, Butte, MT: NCAT (National Center for Appropriate Technology).

Dobson, A. (2000), 'Drei Konzepte ökologischer Nachhaltigkeit', *Natur und Kultur*, **1**: 62–85.

Dorward, A., J. Kydd, J. Morrison and C. Poulton (2005), 'Institutions, markets and economic coordina-tion: linking development policy to theory and praxis', *Development and Change*, **36**(1): 1–25.

Dyllick, T. and K. Hockerts (2002), 'Beyond the business case for corporate sustainability', *Business Strategy and the Environment*, **11**: 130–41.

Easterly, W. (2008), 'Design and reform of institutions in LDCs and transition economies', *American Economic Review: Papers and Proceedings*, **98**(2): 95–9.

Edwards, M. (2008), *Just Another Emperor? The Myths and Realities of Philanthrocapitalism*, New York: Demos.

Gary, I. and T.L. Karl (2003), 'Bottom of the barrel: Africa's oil boom and the poor', Catholic Relief Services Report.

Ghalib, K.A. and F. Hossain (2008), 'Social business enterprises – maximising social benefits or maximising profits? The case of Grameen–Danone Foods Limited', BWPI (Brooks World Poverty Institute) Working Paper No. 51, University of Manchester.

Hackett, M. (2009), '"Social enterprise" in a global financial crisis: is there a developing world voice?' APSA (Australasian Political Studies Association) Conference Sidney, Australia.

Hammond, A.L., W.J. Kramer, J. Tran, R. Katz and C. Walker (2007), 'The next 4 billion: market size and business strategy at the base of the pyramid', Washington, DC: World Resource Institute.

Hart, S.L. (2007), *Capitalism at the Crossroads*, 2nd edn, Upper Saddle River, NJ: Prentice Hall.

Hart, S.L. and M.B. Milstein (2003), 'Creating sustainable value', *Academy of Management Executive*, **17**(2): 56–67.

Hemmati, M., F. Dodds, J. Enayati and J. McHarry (2002), *Multi-Stakeholder Processes for Governance and Sustainability: Beyond Deadlock and Conflict*, London: Earthscan.

Hermoso de Mendoza, A.G. (2011), 'Green management, motor de la Economía Verde', *Sostenibilidad y Conocimiento*, **1**: 6–7.

Holt, D. (2011), 'Where are they now? Tracking the longitudinal evolution of environmental businesses from the 1990s', *Business Strategy and the Environment*, **20**(4): 238–50.

Holt, D. and D. Littlewood (2011), 'Developing a taxonomy of social and environmental enterprises in Sub-Saharan Africa – an initial analysis', Conference on The Business of Social and Environmental Innovation, 14–16 November, Cape Town.

Hopkins, R. (2008), *The Transition Handbook. From Oil Dependency to Local Resilience*, White River Junction, VT: Chelsea Green Publishing.

Ivanko, J.D. (2008), *ECOpreneuring: Putting Purpose and the Planet before Profits*, Canada: New Society Publishers.

Inoue, C.Y.A. and G. Do Prado Lima (2007), *Brazilian Experiences in Sustainable Reserves*, Brasilia: Conservação Internacional.

IUCN, UNEP and WWF (1980), *World Conservation Strategy. Living Resource*

Conservation for Sustainable Development, Gland, Switzerland: Prepared by the International Union for Conservation of Nature (IUCN).

Jeanrenaud, S. (2002), *People-Oriented Approaches in Global Conservation: Is the Leopard Changing its Spots?* International Institute for Environment and Development (IIED) and Brighton: Institute for Development Studies (IDS), London.

Jeppesen, S. and P. Wad (2006), 'Development strategy, industrial policy and cross-border inter-firm linkages', in M.W. Hansen and H. Schaumburg-Müller (eds), *Transnational Corporations and local Firms in Developing Countries: Linkages and Upgrading*, Copenhagen: Copenhagen Business School Press, pp. 311–38.

Katerere, Y. and J.C. Mohamed-Katerere (2005), 'From poverty to prosperity: harnessing the wealth of Africa's forests. Forests in the global balance – changing paradigms', *IUFRO (International Union of Forest Research Organizations, Helsinki) World Series*, **17**: 185–208.

Karnani, A. (2005), 'Misfortune at the bottom of the pyramid', *Greener Management International*, **51**: 99–110.

Leitner, A., W. Wehrmeyer and C. France (2010), 'The impact of regulation and policy on radical eco-innovation', *Management Research Review*, **33**: 1022–41.

Louche, C. and S. Lydenberg (2011), *Dilemmas in Responsible Investment*, Greenleaf Publishing.

Makita, R. (2009), 'New NGO–elite relations in business development for the poor in rural Bangladesh', *Voluntas*, **20**: 50–70.

Masurel, E. (2007), 'Why SMEs invest in environmental measures: sustainability evidence from small and medium enterprises', *Business Strategy and the Environment*, **16**(3): 190–201.

McMullen, J.S. (2011), 'Delineating the domain of development entrepreneurship: a market-based approach to facilitating inclusive economic growth', *Entrepreneurship: Theory and Practice*, **35**(1): 185.

Molteni, M. and A.G. Masi (2009), 'Social entrepreneurship in developing countries: Green technology implementation to push local social and economic innovation', 2nd EMES International Conference on Social Enterprise, Trento (Italy), 1–4 July.

McCormick, J. (1992), *Reclaiming Paradise: The Global Environmental Movement*, 2nd edn, New York: John Wiley & Sons.

Menon, A. and A. Menon (1997), 'Enviropreneurial marketing strategy: the emergence of corporate environmentalism as marketing strategy', *Journal of Marketing*, **61**: 51–67.

Mowforth, M. and I. Munt (1998), *Tourism and Sustainability: New Tourism in the Third World*, London, UK and New York, USA: Taylor & Francis.

Mutagwaba, B. (2009), 'Government expenditure and income inequality in Tanzania: a policy dimension', in S. Sigué (ed.), *Repositioning African Business and Development for the 21st Century*, Proceedings of the 10th International Academy of African Business and Development Annual Conference, Kampala, Uganda, pp. 45–50.

Neumayer, E. (1999), *Weak versus Strong Sustainability: Exploring the Limits of Two Opposing Paradigms*, Cheltenham, UK and Northampton, MA, USA: Edward Elgar.

Noya, A. (ed.) (2009), *The Changing Boundaries of Social Enterprises*, Paris: OECD Publishing.

Obot, A. and I. Obot (2009), 'Sustainable intensive farming and climate change in Africa', in S. Sigué (ed.), *Repositioning African Business and Development for the 21st Century*, Proceedings of the 10th International Academy of African Business and Development Annual Conference, Kampala, Uganda, pp. 45–50.

Osei-Hwedie, K. and A. Bar-on (1999), 'Community driven social policies', in D.A. Morales-Gómez (ed.), *Transnational Social Policies: The New Development Challenges of Globalization*, Ottawa: International Development Research Centre, pp. 89–116.

Oshisanya, K.I. and T.K. Oshisanya (2009), 'Contents of heavy metals of two edible fish *Pseudotolithus Senegalensis* and *Arius heudeloti* from three different geographical locations of Lagos State', in S. Sigué (ed.), *Repositioning African Business and Development for the 21st Century*, Proceedings of the 10th International Academy of African Business and Development Annual Conference, Kampala, Uganda, pp. 5–60.

Pauli, G. (2000), *Upsizing: The Road to Zero Emissions*, Sheffield: Greenleaf Publishing.

Pearce, D.W. and G.D. Atkinson (1993), 'Capital theory and the measurement of sustainable development: an indicator of "weak" sustainability', *Ecological Economics*, **8**: 103–8.

Perrons, D. (2004), *Globalization and Social Change: People and Places in a Divided World*, London: Routledge.

Piebalgs, A. (2007), *Renewable Energy: Potential and Benefits for Developing Countries*, Proceedings of a conference organized by the European Office of the Konrad-Adenauer-Stiftung and the EastWest Institute, Brussels.

Pinchot, G. (1985), *Intrapreneuring: Why You Don't Have To Leave the Corporation To Become an Entrepreneur*, New York: Harper & Row.

Ponnekanti, J. (2011), 'If you teach villagers to make art . . .', *News Tribune*.

Prahalad, C.K. and S.L. Hart (2002), 'The fortune at the bottom of the pyramid', *Strategy and Business*, **26**: 2–14.

Prahalad, C.K. (2009), *The Fortune at the Bottom of the Pyramid*, 5th anniversary edn, Upper Saddle River, NJ: Wharton School Publishing.

Randjelovic, J., A.R. O'Rourke and R.J. Orsato (2003), 'The emergence of green venture capital', *Business Strategy and the Environment*, **12**(4): 240–53.

Rugraff, E., A. Sumner and D. Sánchez-Ancochea (2009), *Transnational Corporations Development Policy: Critical Perspectives*, Basingstoke: Palgrave Macmillan.

Sanya, T. (2007), *Living in Earth. The Sustainability of Earth Architecture in Uganda*, Oslo: Oslo School of Architecture and Design, Oslo.

Schaltegger, S. (2005), 'A framework and typology of ecopreneurship: leading bioneers and environmental managers to ecopreneurship', in M. Schaper (ed.), *Making Ecopreneurs: Developing Sustainable Entrepreneurship*, Burlington, VT: Ashgate, pp. 43–60.

Schaper, M. (ed.) (2005), *Making Ecopreneurs: Developing Sustainable Entrepreneurship*, Burlington, VT: Ashgate.

Scherhorn, G. (2004), 'Sustainability reinvented', Cultures of Consumption Working Paper Series No. 15, London.

Schieffer, A. and R. Lessem (2009), 'Beyond social and private enterprise: towards the integrated enterprise', *Transition Studies Review*, **15**(4): 713–25.

Sehgal, K. (2010), *Walk in My Shoes: Conversations between a Civil Rights Legend and his Godson*, London: Palgrave Macmillan.

Silitshena, R.M.K. (1992), *Botswana: A Physical, Social and Economic Geography*, Gaborone: Longman Botswana.
Simanis, E. and S.L. Hart (2008), *The Base of the Pyramid Protocol: Toward Next Generation BoP Strategy*, New York: Cornell University.
Sparkes, R. (2002), *Socially Responsible Investment: A Global Revolution*, London: John Wiley.
Spring, A. (2009), 'Chinese development aid and agribusiness entrepreneurs in Africa', in S. Sigué (ed.), *Repositioning African Business and Development for the 21st Century*, Proceedings of the 10th International Academy of African Business and Development Annual Conference, Kampala, Uganda, pp. 23–33.
UNCTAD (2007), *World Investment Report Development*.
UNCTAD (2010), *World Investment Report: Investing in a Low-Carbon Economy*.
Von Weizsäcker, E.U., A.B. Lovins and L.H. Lovins (1997), *Factor four. Doubling wealth – halving resource use*.
Williams, A., T. Bonney and M. Xuereb (2009), *Assess the Influence of China in Sub-Saharan Africa*, London: Royal College of Defence Studies.
Wüstenhagen, R., J. Hamschmidt, S. Sharma and M. Starik (eds) (2008), *Sustainable Innovation and Entrepreneurship*, Cheltenham, UK and Northampton, MA, USA: Edward Elgar.
Yunus, M. (2010), *Building Social Business, The New Kind of Capitalism that Serves Humanity's Most Pressing Needs*, New York: Public Affairs Books.
Zahra, S.A., E. Gedajlovic, D.O. Neubaum and J.M. Shulman (2009), 'A typology of social entrepreneurs: motives, search processes and ethical challenges', *Journal of Business Venturing*, **24**: 519–32.

Web References

ACI – The African Cashew initiative, 'Promoting competitiveness of African cashew farmers', available at http://www.africancashewinitiative.org/files/files/downloads/fsaci_promocomp_e150.pdf (accessed 4 June 2012).
Alter, S.K. (2007), 'Social enterprise typology', Virtue Ventures LLC, available at http://www.virtueventures.com/resources/setypology (accessed 4 June 2012).
Brown, K. (2011), 'Sime Darby eyes palm oil expansion in Africa', *Financial Times*, available at http://www.ft.com/intl/cms/s/0/290c6a52-42ad-11e0-8b34-00144feabdc0.html#axzz1wsCs1caZ (accessed 04 June 2012).
Callard, A. (2011), 'Latin America's social enterprises posed for flight', available at http://beyondprofit.com/socents-poised-for-flight (accessed 4 June 2012).
Das Gupta, J. (2011), 'Danone's dual focus on business progress and social progress', available at http://thegreentake.wordpress.com/2011/01/31/danone/ (accessed 4 June 2012).
Elsen, T. (2011), 'Environmental entrepreneurs: Mexico's "Échale a Tu Casa" builds green houses for low-income families', available at http://www.wri.org/stories/2011/04/environmental-entrepreneurs-mexicos-echale-tu-casa-builds-green-houses-low-income-fa (accessed 04 June 2012).
Kennedy, T.F. (2011), 'African elites bear huge responsibility', *Global Geopolitics and Political Economy, Geopolitical and Economic News and Analysis*, available at http://globalgeopolitics.net/wordpress/2011/01/19/african-elites-bear-huge-responsibility/ (accessed 4 June 2012).

Kirkby, C.A.R. Giudice-Granados, B. Day, K. Turner, L.M. Velarde-Andrade, A. Dueñas-Dueñas, J.C. Lara-Rivas and D.W. Yu (2010), 'The market triumph of ecotourism: an economic investigation of the private and social benefits of competing land uses in the Peruvian Amazon', available at http://www.ncbi. nlm.nih.gov/pmc/articles/PMC2947509/ (accessed 4 June 2012).

Masi, A. (2009), 'WISEs in developing countries: the role of the incubator NGO in the management systems', available at http://www.cesvi.eu/UserFiles/File/ paper_WISEs.pdf (accessed 4 June 2012).

Nonprofit Good Practice Guide 2009 (2009), available at http://www.npgoodprac- tice.org/good-practice (accessed 4 June 2012).

Ruli, G. and A. Hoxha (2001), 'Social business and social exclusion in transition countries: the case of Albania', available at http://www.unicef.org/albania/ socialbusinesssocialexclusion.pdf (accessed 4 June 2012).

Sautman, B. and Y. Hairong (2009), 'Trade, investment, power and the China-in-Africa discourse', *Asia-Pacific Journal*, available at http://japanfocus.org/-Yan-Hairong/3278 (accessed 4 June 2012).

SustainAbility (2003), 'The 21st century NGO: in the market for change', avail- able at http://www.sustainability.com/library/the-21st-century-ngo (accessed 4 June 2012).

UK Social Investment Forum (UKSIF) (2007), 'Sustainable and responsible finance in the UK. celebrating 15 years of UKSIF', available at http://www. uksif.org/cmsfiles /15anniversaryreport.pdf (accessed 2 July 2011).

Wang, J. (2011), 'Upcycling becomes a treasure trove for green business ideas', *Entrepreneur*, available at http://www.entrepreneur.com/article/219310 (acces- sed 4 June 2012).

Young, A.J. (2010), 'Rebuilding Haiti with public purpose capitalism', *Huffington Post*, available at http://www.huffingtonpost.com/amb-andrew-j-young/ rebuilding-haiti-with-pub_b_437634.html (accessed 4 June 2012).

8. Entrepreneurship development at a small scale: a key to sustainable economic development

Sanjay Bhāle and Sudeep Bhāle

INTRODUCTION

Successful new business ventures and economic development do not just happen. They are the result of the combination of favourable environment, sincere efforts and meaningful innovation. A rewarding feature of economic development anywhere in the world, be it a developed nation or a developing one, has been the spread in the concept of entrepreneurship and increased numbers of people following it, resulting in many activities at all entrepreneurship levels: medium, small and micro. Small-scale enterprises prove to contribute greatly and to be highly effective, although unnoticed, in creating a significant impact on social development compared to their large-scale counterparts. There is a growing recognition worldwide that small and medium-sized enterprises (SMEs) have an important role to play in the present context, given their greater resource-use efficiency, capacity for employment generation and technological innovation, promoting inter-sectoral linkages, raising exports and developing entrepreneurial skills.

In the recent economic developments the SMEs sector has emerged as a sector that enhances the growth of highly capable entrepreneurs who not only make profits, but also contribute to a better quality of life for the millions who make the profit possible. 'Creative and entrepreneurial energies are generated by the adoption of exogenously supplied beliefs which in turn produce intense efforts in occupational pursuits and accumulation of productive assets leading to manufacturing of goods and services' (Max Weber, in Sami Uddin, 1989: 30).

Entrepreneurs are a product of the particular social conditions in which they live. These social conditions, the society, shape the personality of individual entrepreneurs. The entrepreneur is a leader more than an innovator. 'He/she looks for ideas and puts them into effect for economic development' (Schumpeter, 1934: 105–34).

INNOVATING AND REALIZING THE IDEA

The terms 'creativity' and 'innovation' are often used to mean same thing, but each has a unique connotation. Creativity is the ability to bring something new into existence. It is a phenomenon that stresses ability and not just the activity of bringing something into existence. Ideas usually evolve through a creative process whereby ingenious people bring them into existence, nurture them and develop them successfully with an undaunted spirit. New ideas are the precious currency of the new economy, but generating them does not have to be a mysterious process. Businesses that constantly innovate have systematized the production and testing of new ideas, and the system can be replicated by practically any organization. A good idea for a new product or business practice is not worth much by itself; it needs to be turned into something that can be tested and, if successful, integrated into the rest of the organizational activities.

So how much of the innovation is based on inspiration, and how much on hard work? If it is mainly the former, then management's role is limited: that is, hire the right people, and get out of their way. If it is largely the latter, management must play a more vigorous role: that is, establish the right roles and processes, set clear goals and relevant measures, and review progress at every step.

Peter Drucker, with the masterly subtlety that is his trademark, comes down somewhere in the middle. He says: 'Innovation is real work, and it can and should be managed like any other corporate function. But that does not mean it is the same as other business activities' (Drucker, 2002 [1985]: 2).

Indeed, innovation is the work of knowing rather than doing. It is much easier and safer for organizations to stay with the familiar than to explore the unknown. However in today's fast-changing world, staying with the familiar may have its dangers. The spirit of entrepreneurship, by which creative people are encouraged to come up with new products or services, may become important to the financial health of the organization as well as to social development.

One of the most important aspects of transformation is the total employee involvement in the entire process. In an organization that is being transformed from an 'entrepreneuring organization' to a 'quality management organization', the real change comes from the understanding of the reasons for the success and the failures, and the actions that are taken to bridge the gap.

ENTREPRENEURSHIP: AN ATTITUDE

An entrepreneur always remains an entrepreneur. In fact it is a mindset. The entrepreneur's greatest asset is his adaptability, that is, his ability to keep innovating, understanding the environment and applying concepts in practice.

Drucker argues that most innovative business ideas come from methodically analysing several areas of opportunity, some of which lie within a person, or particular companies or industries, and some of which lie in the broader social or demographic trends. Once one has identified an attractive opportunity, it still needs a leap of imagination to arrive at the right response – call it 'functional inspiration'. Innovation is the specific function of entrepreneurship, whether in an existing business, a public service institution, or a new venture started by a lone individual in the family kitchen. (There are some small home-based businesses run by individuals providing packed meals on regular basis to working class households, which are not only cheaper than what is available on the market, but also cater to a certain requirement of the households.) It is the means by which the entrepreneur either creates new wealth-producing resources or endows existing resources with enhanced potential for creating wealth; hence the question for a would-be entrepreneur is: how can one translate innovation into an operational organization (an operational organization means bringing the innovation to a practical means to provide a regular and effective source of income)? To help discover answers, we must first look at entrepreneurial behaviour. And such behaviour is developed by social motivation with the help of certain paradigms upon which microeconomies work.

Entrepreneurial behaviour generates a guiding spirit for the rest of the society and thus inculcates the spirit of collaborative social networking, innovation and continuous improvement, steadfastness in achieving the objectives and a noble vision worth emulating. Being entrepreneurial is all about leveraging mutually beneficial relationships and nurturing hidden talents of the team. It covers all the aspects of leadership, as an enterprise endowed with the resource of leadership always stays consistent in the market. Entrepreneurship essentially is a noble act of sustainable development. Though the idea upon which any entrepreneurial venture is based may be small, it is very important, as it is an innovation which leads to entrepreneurship.

One of the relatively new terms evolved in recent times linked to entrepreneurship is 'social entrepreneurship', which connects the fundamental aim of wealth creation to a greater degree of sustainability for small-scale entrepreneurial developments. The basis of such developments is: for society, by society and within society to achieve more sustainable continuous economic growth. Below, we will explore social entrepreneurship further.

EMERGENCE OF 'SOCIAL ENTREPRENEURSHIP'

The phrase 'social entrepreneurship' was coined in the late 1990s to describe a business practice. Subsequently, academics debated the concept (Johnson, 2000). Traditionally, only commercial or business entrepreneurship was recognized. However, the empowerment and awareness among the less privileged communities, especially in the developing world, has resulted in the increased growth of social entrepreneurship. Advances in communication technologies have increased everyone's access to information, providing knowledge that individuals can apply to exploit opportunities to meet their social needs. Apart from empowering the communities, advances in communication and technologies have provided new ways for social entrepreneurship to address social issues globally, especially in the developing world. Several key factors are fuelling social entrepreneurship in the global context (Zahra et al., 2008). Global wealth disparity and the corporate social responsibility movement are sensitizing the developing countries to explore opportunities for social improvement. Failure of markets or that of the government to meet the basic social needs of the citizens, especially those at the bottom of the pyramid (BOP), has led to emergence of social entrepreneurship. Social entrepreneurship is built on the concepts of traditional entrepreneurship. However, there are several similarities and several differences between business entrepreneurship and social entrepreneurship.

BUSINESS VERSUS SOCIAL ENTREPRENEURSHIP

A dynamic theory of entrepreneurship was first advocated wherein it acts as a catalyst that disrupts the stationary circular flow of the economy, and thus initiates and sustains the process of development (Schumpeter, 1990 [1934]). Formerly, entrepreneurship was considered as any purposeful activity that initiates, maintains or develops profit-oriented business in interaction with the internal situation of the business or with the economic or social circumstances surrounding the business. This approach emphasized the coordination activity and sensitivity to the environment affecting decision-making (Cole, 1995 [1949]). The essence of 'traditional' entrepreneurship was the alertness of market participants to profit-making opportunities (Kirzner, 1973). The broader definition of business entrepreneurship which emerged over a period of time was that it is: 'a force that mobilizes other resources to meet unmet market demands', 'the ability to create and build something from practically nothing', and 'the process of creating value by pulling together a unique package of resources to exploit an opportunity'.

A succinct definition in use today is that entrepreneurship is the process of pursuing limitless opportunities using the resources currently in hand, with an emphasis on the word 'process'. This process of entrepreneurship, in the commercial context, is considered to consist of five components (Morris, 1998: 29), that is: opportunity recognition, concept development, resource determination and acquisition, launch and venture growth, and harvesting the venture. Similarly, social entrepreneurship is viewed as a process of creating value by combining resources in innovative ways (Mair and Marti, 2006). It is a process that catalyses social change and addresses important social needs in a way that is not dominated by direct financial benefits for the entrepreneurs. This resource mobilization is intended primarily to create social value to meet social needs.

Opportunity recognition and innovation are the core components of entrepreneurship, which are also key components of social entrepreneurship. Opportunity is a favourable set of circumstances for doing something such as establishing a new venture. This is often in response to a mission statement identification. When a mission is identified, opportunities are taken to innovate to address the mission. Thus opportunities, at the core of entrepreneurship, constitute one of the commonalities between commercial or business entrepreneurship and social entrepreneurship (Drucker, 2002 [1985]). Opportunities are recognized where unmet needs exist, whether these are social needs or economic needs (Mair and Marti, 2006). However, the opportunities for social entrepreneurship are likely to be distinct from opportunities in the commercial sector (Robinson, 2006).

Traditionally, organizations addressing a social problem were considered as idealistic, philanthropic and non-entrepreneurial. However, more recently with merging of the private and social sectors, there has been a realization that pure philanthropic or pure capitalist approaches are not sustainable. Social entrepreneurship has evolved from commercial entrepreneurship to bring about a dignified social change or to bring about social equity which is not through traditional approaches such as charity. Many business enterprises with considerable financial gains make charitable gestures for the community through the enterprise. However, this is not a social enterprise. Social enterprise is an enterprise which believes in 'doing charity by doing business' rather than 'doing charity while doing business'. Thus, essentially social entrepreneurship utilizes the concept of commercial entrepreneurship to achieve social goals.

Social entrepreneurship is a much newer concept as compared to commercial entrepreneurship. Social entrepreneurship has many facets and includes a range of innovative, dynamic, social value creating ventures. 'The business of business is business', and it does not address inequity and sustainability. Social entrepreneurship represents an interesting alternative model for business as well as addressing inequity and sustainability on a large scale. Socially entrepreneurial ventures (SEVs) stand out and are unique as they focus on a different set of possibilities: innovative ways to bring about sustainable social change. This is brought about by merging business with non-profitability. Santos (2009) differentiates social entrepreneurship from commercial entrepreneurship on the basis of value creation and value appropriation. Social enterprises differ from commercial enterprises in that they do not aim to offer financial benefit to their investors. The financial gains created are utilized for increasing their capacity to realize the social goals. Thus, the main difference between entrepreneurship in the business sector and social entrepreneurship lies in the relative priority given to social wealth creation versus economic wealth creation (Mair and Marti, 2006).

DEFINITION OF SOCIAL ENTREPRENEURSHIP

There are several definitions of social entrepreneurship and a standard definition is yet to emerge. Social entrepreneurship addresses social issues or unmet needs of the economically less privileged community, hitherto not met by public or private agencies. It provides innovative solutions to social issues through implementing innovative ideas, mobilizing resources and capacities required for sustainable social transformation (Alvord et al., 2004). Social enterprises are considered to be dedicated to solving social problems, serving the disadvantaged and providing socially important goods that were not, in their founders' judgment, adequately provided by public agencies or private markets (Dees, 2005).

Social entrepreneurship is often confused with business models that are designed to create accountability within the business sector. Clarity is required to differentiate between 'social ventures' and 'social enterprises'. Social ventures are the result of social venture capital, whereas social enterprise is built by financing through means other than venture capital. Regardless of the source of funding, the social ventures and social enterprises are a result of social entrepreneurship. Social entrepreneurship also needs to be differentiated from terms such as 'sustainable enterprise', 'corporate social responsibility' and 'business ethics' (Schaeffer and Craig, 2007). Both social enterprises and social ventures resulting from social entrepreneurship may practice social responsibility or sustainability; and yet they are different from a sustainable enterprise or corporate social responsibility. Social entrepreneurship as part of the social responsibility of business organizations has gained popularity. Several venture capital firms have invested in for-profit entities with social objectives. These venture funds measure their investments on social, environmental and the traditional financial returns.

Social entrepreneurship is motivated primarily by social benefit. It is a multidimensional construct involving the expression of entrepreneurially virtuous behaviour to achieve a social mission (Mort et al., 2003). Social entrepreneurship is not about balancing the:

> double bottom lines of profit and social impact. The real impact that a social enterprise should show is on social or environmental impact. It should address a social mission through utilizing the financial returns earned. It is not a fundraising strategy for social organizations. Funds generation is not the primary aim of the enterprise, the primary and the only aim is a sustainable social impact. Social entrepreneurship is broadly viewed as a process involving the innovative use and combination of resources to pursue opportunities to catalyse social change and/or address social needs. (Mair and Marti, 2006: 36–44)

Efforts are being made by researchers to arrive at a clear definition of social entrepreneurship. A clear distinction is made between social entrepreneurship and entrepreneurship in general. However, research is required to identify the variables or parameters of social entrepreneurship, as they vary based on the region in which they are implemented. This regional diversity makes it difficult to define it as a global phenomenon. It needs to be defined in general terms in reference to the fundamental goals it is supposed to achieve.

KEY FEATURES OF SOCIAL ENTREPRENEURSHIP

Martin and Osberg (2007) define social entrepreneurship as consisting of three components:

Identifying a stable but inherently unjust equilibrium that causes the exclusion, marginalization or sufferings of a segment of humanity that lacks the financial means or political clout to achieve any transformative benefit on its own;

Identifying an opportunity in this unjust equilibrium, developing a social value proposition, and bringing to bear inspiration, creativity, direct action, courage and fortitude, thereby challenging the stable state's hegemony; and

Forging a new, stable equilibrium that releases trapped potential or eliminates the suffering of the targeted group, and through innovation and the creation of a stable ecosystem around the new equilibrium ensuring a better future for the targeted group and society at large.

Another major feature of social entrepreneurship is that it is the result of vision-oriented and crisis-oriented factors, like traditional entrepreneurship (Mair and Marti, 2006). Social entrepreneurship and its dual-value objective can serve as an interesting model for traditional corporations to fulfil increasing societal expectations (Davidsson and Honig, 2003). There is a consensus that for-profit activity is necessary for achieving sustainable solutions to poverty at the so-called bottom of the pyramid (BOP), the poorest segment of a society. Social enterprises generally work with market forces and not against them so that there are financial gains along with social gains.

Social entrepreneurship is not about balancing profit and social impact. Profits are achieved along with addressing a social issue. Profit-making is not the primary goal; it is a means of attaining a social change. Social entrepreneurship is about achieving a social mission; and the financial benefits of the enterprise are just a means to achieve the ultimate goal, that is, social change. This is not the exclusive domain of non-profits. Globally, non-profit organizations bring about commendable social change. However, they are not considered as social enterprises as they depend on governments, philanthropists and other donors. On the other hand, simply being part of a profit-making organization does not make it a social enterprise. It has been realized that continued separation of the social and the economic factors is strategically unsustainable for big businesses (Nicholls, 2006). Good businesses understand that a proactive reduction of inequity and trust crisis is a good business strategy. Thus, a social enterprise is any for-profit or non-profit enterprise that applies capitalistic strategies to achieving philanthropic goals.

Duncan (2007) mentions the key variables such as motivation, capacity and start-up capital in the entrepreneurial non-profit model. La Pira and Gillin (2006) identified macro-level reinvention or reengineering; a good public policy; reconfiguration and taxation; and value creation through transcendence as the key parameters of social entrepreneurship which creates value for the society.

Social entrepreneurship plays an important role in bringing 'sustainable' solutions in addressing poverty and thus bridging the divide between rich and poor (Drucker, 2002 [1985]). Sustainability is another key feature of social entrepreneurship (Santos, 2009). Empowerment of the community ensures sustainability in an enterprise. A social enterprise maintains a balance between value creation and value appropriation. An enterprise generates profits, which makes people accountable and motivates further growth. The profits arc reinvested for further development, ensuring the sustainability of an enterprise. Social wealth creation is the primary objective of a social enterprise, and economic value creation, in the form of earned income, is a necessary commodity to ensure the sustainability of the enterprise and financial self-sufficiency (Mair and Marti, 2006). Social entrepreneurship captures the elements from traditional entrepreneurship as profitability makes a big impact on the success of an enterprise and reinventing the business models to use the cycle of profitability in creating sustainable developments (Schramm, 2010). Sustainability and scalability are important parameters to judge the impact of a social enterprise.

A literature review on addressing the BOP reveals that any activity that generates revenue by selling products or procuring products from the people at the BOP in such a way that it improves their standard of living is considered a BOP venture. Prof. C.K. Prahalad (2005) believes that major companies have not yet targeted the poor people as potential consumers by innovating low-cost consumer products. This market of potential consumers, estimated to be around 5 billion globally who earn less than $2 per day, can be reached by creative entrepreneurs who are prepared to offer affordable products and services that meet the needs of the poor. This is a sure way in which both the entrepreneur and the poor are benefited immensely, thus creating wealth and reducing poverty. Today, several ventures claim to be social enterprises professing poverty alleviation. These kinds of assertions are not scrutinized as there is no precise set of structure to validate their claims that poverty alleviation was their primary goal and not profit-making. The essential criteria in deciding whether a social entrepreneur is contributing towards a social impact such as poverty reduction is whether the venture is directly involved in serving the poor. Thus, it is believed that a key characteristic of social entrepreneurship is involving the people at the BOP for ensuring social equity.

People's involvement in the entire entrepreneurial process is an important characteristic of this type of entrepreneurship. Social entrepreneurship actively involves several stakeholders from the representative community or society to ensure their direct inputs in the operation of the enterprise. It is often a case of collective action and multiple actors working together to create social value (Corner and Ho, 2010). Thus, social enterprises aiming at economic equity target an underserved or highly disadvantaged population, that is, those at the bottom of the pyramid (BOP), involving them in the enterprise. One excellent example of involvement of this section of population for their own improvement is the Grameen Bank of Muhammad Yunus (Mohammed, 2008). The Grameen Bank has been successful in uplifting the rural poor in Bangladesh by developing a revolutionary micro-credit system that is cost-effective and scalable to fight poverty. The impact of social entrepreneurship lies in its successful implementation through the people at the BOP, the most vulnerable. It is a scalable, sustainable enterprise based on the needs of the poor or less privileged, wherein the communities themselves are involved for their own good.

Another example cited of such an enterprise is the access to Wikipedia and other such facilities, which is very difficult in several parts of the world. Rodrigo Baggio, founder of the Committee for Democracy in Information Technology in Brazil, addressed this through social entrepreneurship (Ashoka Innovators for the Public, n.d.). The youth of the slum area of Rio De Janeiro in Brazil were given access to the Internet, thus eliminating their digital exclusion. The young people of the region were empowered with skills of using information technology to explore employment opportunities to bring about economic uplift of the community. This is an excellent example of a sustainable solution to bring about economical equity among the affected masses.

The Grameen Bank and other such endeavours fit into the definition of social entrepreneurship wherein the social interests combine with business practices to effect social change (Peter et al., 2006). These examples emphasize that social value creation is the primary goal and that economic value creation is considered a necessary condition to ensure financial viability of the enterprise. However, initiating a new venture, as in case of the Grameen Bank, is only one type of social entrepreneurship. Mair and Marti (2006) emphasize that social value creation is either offering new services or products, or can also involve the creation of new organizations. Several other categories of possible initiatives can be considered as a social enterprise. Six possible categories are listed by Brinckerhoff (2000: 16–21):

- starting a new product or service;
- an existing product or service;
- expanding an existing activity to a new group of people;
- expanding an existing activity to a new geographic area;
- acquiring an existing product or service;
- partnerships with an existing service or business.

Each of the above categories can be classified as a social enterprise. The bottom line is that a new opportunity is recognized to provide a useful social service.

The belief that addressing a social issue is a moral obligation that cannot be fulfilled by means of business, no longer holds true. The paradigm shift took place when it was realized that businesses can generate profits from these untapped markets and simultaneously provide a better way of life to the citizens of the society. Social entrepreneurship has evolved from philanthropy to non-profit to self-sustainability, and it will keep evolving based on social needs. Social entrepreneurship, like business entrepreneurship, should not be viewed purely in the economic sense; it needs to be examined in light of the social context and the local environment (Mair and Marti, 2006).

MILLENNIUM DEVELOPMENT GOALS: THEMES AND TARGETS

Currently, there are various social sectors in which social entrepreneurship invests, such as education, health, rural development and environment. Education as a sector has already seen social enterprises emerge, as many of these ventures have not only reached break-even but some of them have emerged as profit-making, increasing their sustainability factor. The health and rural development sectors have great potential for the future growth of social enterprises. Considering the above characteristics of social entrepreneurship, it can be used effectively to realize the Millennium Development Goals (MDGs) signed by the United Nations for adoption of a global action plan to achieve the eight anti-poverty goals (UNDP, n.d.). MDGs can be achieved through innovative and sustainable ideas which are the key of social entrepreneurship. Entrepreneurs need to bring about social equity in the world (Brinkerhoff et al., 2007). Social entrepreneurship will not only achieve the MDGs but also create sustainable, profitable business models.

A MODEST TESTIMONY

Here is a classic case that exhibits the unequivocal relationship of entrepreneurship with sustainable development. A small organization run by women for women started in 1959 with seven lady members and a borrowed sum of Rs 80/- in Mumbai, India. It was a small group of illiterate housewives with lots of time but no resources, spending the day gossiping in their cramped, lower-middle-class dwellings till they hit upon the idea of selling *papads* (India's most popular crispy bread). Initially, getting a foothold was a lot of struggle. However in the period to come they succeeded on the basis of sheer honesty and hard work. This resulted in setting new standards in the market and also in generating employment at the local level. Fifty years later, their entrepreneurial effort, Lijjat Papad (see Lijjat Papad Organization, n.d.), has turned into a cooperative, Shri Mahila Griha Udyog Lijjat Papad (SMGULP).This is a women's organization which symbolizes the strength of women; only women can become members of the organization and usually they are called 'sister' or *ben*. The organization is extensive, with its central office at Mumbai and 72 branches and around 35 divisions in different states all over India. Today, Lijjat is more than just a household name for *papad*. Encouraged by the popularity of the brand name, it has diversified into spices, wheat flour, soaps, detergents and snacks.

The co-operative is now an award winning company with an annual turnover of more than Rs 6.5 billion (~US$130 million, year 2010). Everyone enjoys 'rags to riches' stories in a modest way and everyone likes tales of stupendous success achieved through sheer determination. Over 42000 members throughout India have withstood hardships with unshakable belief. And perhaps that is the most interesting lesson managers can learn from such an organization. Sticking to its core values for the past 50 years, Lijjat has ensured that every process runs smoothly, members earn a comfortable profit, agents get their due share, consumers get the assurance of quality at a good price, and the society benefits from its donations to various causes.

SMGULP is a synthesis of three different concepts, namely: (1) the concept of business; (2) the concept of family; (3) the concept of devotion. These concepts are completely and uniformly followed in this institution. The institution has adopted the concept of business from the very beginning. All its dealings are carried out on a sound and pragmatic footing: production of quality goods and selling at reasonable prices. It does not accept any charity, donation, gift or grant from any quarter. On the contrary, the members donate collectively to good causes from time to time according to their capacity. Besides the concept of business, the institution along with all its members has adopted the concept of mutual family affection, concern and trust. All affairs of the institution are dealt in a manner similar to that of a family traditionally carrying out its own daily household chores.

But the most important concept adopted by the institution is the concept of devotion. For the member sisters, employees and people who are associated with organization in any way, the institution is never merely a place to earn one's livelihood. It is a place of worship to devote one's energy not for individual benefits but for the benefit of all. In this institution, work is worshipped. The institution is open for everybody who has faith in its basic concepts.

What is more stunning than its stupendous success is its striking simplicity. It is a story that shows how an organization can infuse Gandhian simplicity in all its activities. (Gandhi believed in a life of simplicity and self-sufficiency, simple living encompasses a number of different voluntary practices to simplify one's lifestyle. These may include reducing one's possessions or increasing self-sufficiency.)

THE BUSINESS MODEL

The entire cycle starts with a simple recruitment process. Any woman who pledges to adopt the institution's values and who has respect for quality can become a member and co-owner of the organization. In addition, those involved in the production also need to have a clean house and space to dry the *papads* they roll every day. Those who do not have this facility can take up any other responsibility, such as kneading dough or packaging or testing for quality. If we look at its distribution cycle it is simply organized. Every morning a group of women goes to the Lijjat branch to knead dough, which is then collected by other women who roll it into *papads*. When these women go to collect the dough, they also give in the previous day's production, which is tested for quality.

Packed *papads* are sealed into a box (each box holds 13.6 kg) and the production from each centre is transported to the depot for that area. Each depot stocks production from the nearby three to four branches, roughly about 400 boxes. In some smaller towns or villages, the branch itself serves as the depot. The depots are storage areas as well as pick-up points for distributors. Some part of the production is also exported. Only upon receiving the full advance through a cheque does production begin. Because all exports are made from Mumbai, the supply also comes from there. The exported production is of the same quality as the daily production. Around 30 per cent of the production of Lijjat Papad is being exported, mainly to countries such as the United States, the United Kingdom, the Middle East, Singapore, Hong Kong and Holland.

The organization works on empowering everybody who works with it; the organization has never shied away from sharing power in all its activities. All sister-members of the institution are the owners, all profit or loss is shared. Only they have the authority to decide the manner in which profit or loss should be portioned among them. The committee of 21 members manages the affairs of the institution. There are also supervisors, who look after the daily affairs of a centre. At the same time, each and every member has the power of veto. All decisions, major or minor, mostly are based on consensus among members.

Such an institutional set up helps avoid the usual management nightmares. For instance, production is carried out not in one central location but in hundreds and thousands of individual homes. The branch system ensures that every activity happens within already set guidelines and scope, and to a set standard of quality. Testing for quality and packaging is done at every branch. Certain activities, however, are centralized: for example, all raw materials are purchased in Mumbai and then distributed to various branches to ensure consistent quality. The organization has accountants in every branch and every centre to maintain daily accounts. Profit (or loss, if any) is shared among all the members of that branch.

There is a committee that decides how the profits are to be distributed. Everyone gets an equal share of profit, irrespective of who does what work, irrespective of seniority or responsibility.

RETAINING VALUES

The business is run on certain values: every member who joins it pledges, 'we will make all-round effort to ensure that the *bens* [sister-members] get real fruit of their labour and we will not allow to happen any type of economic loss to the institution knowingly, unknowingly, directly or indirectly'. Every member is aware that it is one of the very important traditions of the institution that neither sister-members nor employees wrongfully take away any money or material from the institution. Among others, the chief value that holds the institution firmly together is a sense of self-dignity and respect. It discourages any kind of class distinction and does not declare this organization to be for poor women. However, though it would make business sense to adopt modern technology for mass production or use machines for packaging, and so on, this has not been done because it would defeat the very purpose of the business's existence, which is to provide a source of livelihood and dignity to women through self-employment.

The beginning was so modest: seven women with no special skills, but a strong determination to earn their living with dignity as individuals, went ahead to make a successful business, doing what they knew best – rolling *papads*. They leveraged their basic skills and turned these into a tool for success because they believed in themselves and in each other. It is this belief that has been the basis of the business model.

Lijjat helps those women who are not encouraged to work outside their homes to contribute to the family income. These women take dough home and roll it into *papads* when they are free from their domestic chores. At the same time it is not prudent for the organization to invest in office property for so many members. The perfect fit for both is using the members' homes for the rolling and drying of *papads*; no additional over-head costs, no investment either. Valuing people and understanding their problems has created a sound and sustainable business mode for Lijjat.

From this case it is very apparent that entrepreneurship is not about money; it is actually following a purpose, a dream, to be focused upon with passion. A lot of people believe that entrepreneurship is about making money or a lifestyle, which may not be the case for every entre-preneur. For some of them it is about creating wealth in the direction of alleviating poverty, and uplifting education and healthcare standards in the society. It is essentially an attempt to provide solutions by adopting new approaches. It is about sensing the opportunities and responding to them in the most sensible way. An idea can be generated only by sheer observation. Though an idea takes time to evolve, it takes great efforts to nurture it, and to convert it into a sensible innovation – an innovation that can be transformed into reality and is sustainable enough to provide empowerment to the society. One may have to live with an idea for a while.

SMALL IS BIG

Recognizing the growing need for and employment importance of MSMEs (micro, small and medium-sized enterprises), the private sector has to be encouraged and supported with innovative applications of developmen-tal schemes. Such efforts eventually facilitate the expansion of markets and new business opportunities. With growing accountability and the development of social and environment standards in the global market, competitiveness gets linked to a host of factors. It is therefore becoming increasingly imperative to encourage the small and medium-scale enter-prises to be transparent in order to attain sustainable development over the long term in the future. Small businesses are not just small versions of big businesses. They have a number of distinctive features of leadership and innovation that are not always obvious or noticed in general practice. According to Prof. C.K. Prahalad (2007) in his book *The Fortune at the Bottom of the Pyramid*, the basic economics of the market are based on small unit packages, low margins per unit, high volumes and high return on capital employed. The bottom of pyramid approach demands a range of innovations in products and services, business models and management processes. These innovations impact upon the life of manufacturers and, particularly, consumers at large. As the consumers get an opportunity to participate in and benefit from the choices of products and services made available through the market mechanism, the accompanying social and economic transformation can open new avenues for future development. The reason for this is that consumers at the bottom are very demanding, and can easily imagine ways in which they can use their newly found access to information in the most beneficial manner by making appropriate choices.

Those who wish to understand small businesses therefore need to develop insight into the specific nature of such organizations. Leadership needs to shift from perpetuating the status quo to being a catalyst of change. With creative leadership, a region can become innovative at the societal level. This in combination with the innovative abilities of individuals and businesses can help societies reach a strategic inflection point that finally puts them on a path that lifts millions to a self-sufficient and empowered life. Businesses are important and vital since they provide employment opportunities and potential for growth. We should use the engines of economic growth and entrepreneurship to create an attitude of self-reliance. Sustainable economic development is organically related to the impulse of change, reflecting the developmental urge of the society for self-reliance and willingness to observe painstaking discipline. It needs to create a business environment and a social culture that will let the millions of budding entrepreneurs turn their dreams into reality.

REFERENCES

Alvord, S.H., L.D. Brown and C.W. Letts (2004), 'Social entrepreneurship and societal transformation: an exploratory study', *Journal of Applied Behavioral Science*, **40**(3): 260–282.

Brinckerhoff, P.C. (2000), *Social Entrepreneurship: The Arts of Mission-Based Venture Development*, New York: Wiley, pp. 16–21.

Brinkerhoff, J.M., S.C. Smithand and H. Teegen (eds) (2007), *NGOs and the Millenium Development Goals: Citizen Action to Reduce Poverty*, New York: Palgrave Macmillian.

Cole, A.H. (1995 [1949]), 'Entrepreneurship and entrepreneurial history', in H.C. Livesay (ed.), *Entrepreneurship and the Growth of Firms*, Vol. I, Aldershot, UK and Brookfield, VT, USA: Edward Elgar, pp. 100–122.

Corner, P.D. and M. Ho (2010), 'How opportunities develop in social entrepreneurship', *Entrepreneurship Theory and Practice*, **34**: 635–59.

Davidsson, P. and B. Honig (2003), 'The role of social and human capital among nascent entrepreneurs', *Journal of Business Venturing*, **18**(3): 301–31.

Dees, J.G. (2005), 'Social Entrepreneurs and Education', *Current Issues in Comparative Education*, **8**(1): 51–5.

Drucker, P.F. (1994), 'Our entrepreneurial economy', *Harvard Business Review*, **62**(1): 58–64.

Drucker, P.F. (2002 [1985]), 'The discipline of innovation', *Harvard Business Review*, reprint, Boston, MA: Harvard Business School Publishing Corporation.

Johnson, S. (2000), 'Literature review on social entrepreneurship', Canadian Centre for Social Entrepreneurship.

Kirzner, I.M. (1973), *Competition and Entrepreneurship*, Chicago, IL: University of Chicago Press.

La Pira, F. and M. Gillin (2006), 'Non-local intuition and the performance of serial entrepreneurs', *International Journal of Entrepreneurship and Small Business*, 3(1): 17–35.

Martin, R.L. and S. Osberg (2007), 'Social entrepreneurship: the case for definition', *Stanford Social Innovation Review*, Spring: 27–39.

Mohammed, O. (2008), 'Role of microfinance in poverty alleviation: lessons from experiences in selected IDB member countries', Social Science Research Network.

Morris, M.H. (1998), *Entrepreneurial Intensity: Sustainable Advantages for Individuals, Organizations, and Societies*, Westport, CT: Quorum Books.

Mort, G.S., J. Weerawardena and K. Carnegi (2003), 'Social entrepreneurship: towards conceptualization', *International Journal of Non-profit and Voluntary Sector Marketing*, 8(1): 76–88.

Nicholls, A. (ed.) (2006), *Social Entrepreneurship: New Models of Sustainable Social Change*, New York: Oxford University Press.

Peter, R., K. Sarath and B.M. Wasma (2006), 'Reassessing necessity entrepreneurship in developing countries', Institute for Small Business and Entrepreneurship (ISBE).

Prahalad, C.K. (2005), *The Fortune at the Bottom of Pyramid*, Upper Saddle River, NJ: Wharton School Publishing.

Robinson, J. (2006), 'Navigating social and institutional barriers to markets: how social entrepreneurs identify and evaluate opportunities', in J. Mair, J. Robinson and K. Hockets (eds), *Social Entrepreneurship*, New York: Palgrave Macmillan, pp. 95–120.

Schramm, C. (2010), 'All entrepreneurship is social', *Stanford Social Innovation Review*, Spring: 20–22.

Schumpeter, J.A. (1990 [1934]), 'The theory of economic development', in M. Casson (ed.), *Entrepreneurship*, Cheltenham, UK and Northampton, MA, USA: Edward Elgar, pp. 105–34.

Uddin, Sami (1989), *Entrepreneurship Development In India*, Delhi: Mittal Publications.

Web References

Ashoka Innovators for the Public (n.d.), Research Fellow Rodrigo Baggio Barreto, available at http://www.ashoka.org/node/3396 (accessed 20 June 2011).

Lijjat Papad Organization (n.d.), available at www.lijjat.com (accessed 20 April 2012).

Mair, J. and I. Marti (2006), 'Social entrepreneurship research: a source of explanation, prediction, and delight', *Journal of World Business*, 41(1): 36–44; available at www.sciencedirect.com (accessed 20 May 2011).

Quaid, N. (2009), 'Lijjat – empowering women through papads for 50 years' available at http://news.gaeatimes.com/lijjat-empowering-women-through-pap ads-for-50-years-17630/ (accessed 1 June 2011).

Santos, F.M. (2009), 'A positive theory of social entrepreneurship', *Social*

Entrepreneurship e-Journal, Social Science Research Network, INSEAD Paper No. 2009/23/EFE/ISIC (INSEAD Social Innovation Centre), available at www. ssrn.com (accessed 25 April 2011).

Schaeffer, L. and P.D. Craig (2007), 'Social entrepreneurship. Encyclopaedia of business ethics and society', Sage Publications, available at http://www.sage-ereference.com /ethics / article_n755.html (accessed 6 May 2011).

United Nations Development Programme (UNDP) (n.d.), 'Millennium Development Goals', available at http://www.undp.org/mdg (accessed 5 June 2011).

Zahra, S.A., H.N. Rawhouser, N. Bhawe, D.O. Neubaum and J.C. Hayton (2008), 'Globalization of social entrepreneurship opportunities', *Strategic Entrepreneurship Journal*, **2**: 117–31, available at www.interscience.wiley.com (accessed 20 May 2011).

9. Entrepreneur profile and sustainable innovation strategy

Sandrine Berger-Douce and Christophe Schmitt

INTRODUCTION

The environmental challenges (climatic change, limited supply of natural resources, the change in the ozone layer) associated with increasing population create a favorable environment for using innovation opportunities (Abrassart and Aggeri, 2007). In a global context that creates a strong argument for a green economy (Grenelle de l'Environnement in October 2007 in France, windmill parks in Denmark and California), the numerous economic protagonists, particularly small businesses, are often forgotten during debates about sustainable development, unlike the multinationals which are considered to be the experts on 'green-washing' according to non-governmental organizations (NGOs) such as Les Amis de la Terre (Friends of the Earth) and numerous authors (Davis, 1992). At the same time, in an approach inspired by resource theory (Wernerfelt, 1984; Barney, 1991), innovation strategies are recognized as the guarantee for long-term survival of organizations, irrespective of their size (Camison-Zomosa et al., 2004; Soparnot and Stevens, 2007). As well as being a powerful lever for earning acceptability and legitimacy (Mathieu and Reynaud, 2005), strategies for sustainable development, defined as strategies aiming for overall performance including social and environmental aspects, represent real opportunities, particularly economic, for businesses. For McWilliams et al. (2006), CSR (corporate social responsibility, as the managerial translation of sustainable development) is a 'form of investment strategy'. In relation to small businesses, the personal aspirations of a small business leader have a great influence on the business's strategies (Jaouen, 2008).

In this chapter, we would like to discuss the reasons why strategic, technological and innovative choices are made and deployed by economic actors, paying close attention to the relationship between entrepreneurs' profiles and the strategies for sustainable innovations that they develop. Our hypothesis is that the entrepreneur's profile is one of the key factors for success of sustainable development strategies. The entrepreneur's profile model developed by Daval et al. (2002) and used in diverse entre-preneurial situations, such as the development of small innovative busi-nesses (Boissin et al., 2008), is enlightening when applied to other topics such as strategies for sustainable development. The chapter first presents the theoretical framework of research about the relationship between innovation, entrepreneurship, and sustainable development. It then gives a case study of a small business in the north of France and, more specifi-cally, its strategy for sustainable technology and innovation.

STRATEGIES FOR INNOVATION, ENTREPRENEURSHIP AND SUSTAINABLE DEVELOPMENT

In this section the objective is to show the relationship between an entre-preneur's profile and the resulting strategies for sustainable development. To approach this relationship, we first looked at the position of innova-tion in the entrepreneurial process, namely, strategies for sustainable innovation. The final point deals with entrepreneurs' profiles in relation to innovation. What can be concluded from different studies in this subject area is a model that allows us to make a link between the entrepreneur's profile and the strategies for sustainable innovation.

Innovation and Entrepreneurship

According to Julien and Lachance (2006: 335), 'Entrepreneurship is firstly the creation of collective values, recognized by different agents that share the challenges and the risks of innovation accepted by the interior market, secondly the middle market, and finally the exterior'. If we believe in this definition, social strategy is an integral part of entrepreneurship. Beyond the typologies of entrepreneurs found in the existing literature, an entrepreneur combines the characteristics of the primary paradigms of the subject area of entrepreneurship (Verstraete and Fayolle, 2005). The integration of sustainable development in the area of entrepreneurship can thus be seen as the construction and operation of an opportunity in the form of new ideas, such as the creation of a new organization or a new value. The creation of values and the long-term survival of an organiza-tion are the key elements of the general policy of the evidence of Daval et al. (2002) shown in this research.

Innovation is one of the essential factors of an organization's per-formance (Foray and Mairesse, 1999; Raymond and St-Pierre, 2008) and of the national economy (OEDC, 2006). 'Innovation has a strategic dimension. It determines the ability of an organization to beat its competitors in the long run' (Soparnot and Stevens, 2007: 7). For 40 years, significant progress has been made in comprehending innovation; however, there are still certain ambiguities (Carrier and Garand, 1996). According to Tremblay (1997), for microeconomics just as for macro-economics, the relationship between innovation and performance is the object of several works in economic literature since Schumpeter's analysis (1935). Schumpeter conceives innovation as the moment of achievement within a new combination of ideas. Under this general idea of innovation, he groups five different forms: the making of a new item, the introduction of a new production method, a new outlet, the sourcing of raw materials and the creation of a new organization. For Schumpeter, this combination defines a business. The entrepreneur is not an inventor, but someone who will introduce the discovery into the enterprise (and so into the industry). To do this, the entrepreneur must overcome certain objective resistance (linked to the procedures for implementing the innovation), subjective resistance (deviating from a beaten path) and social resistance (coming from clients and competitors) in order for the innovation to be accepted. One of the primary contributions of Schumpeter's research is his 'vision of the entrepreneur as an agent, an innovator and a strategist functioning with his management choices' (Tremblay, 1997: 18). For Van de Ven (1986), innovation corresponds to the development and putting to work of new ideas by people committed to transactions with others in a given organizational context. Kanter (1983) describes innovation as a process leading to the use of new techniques for resolving problems and to the production of new ideas. A characteristic of the evolutionary vision of innovation is that it is considered to be a learning process. 'In other words, history counts!' (Tremblay, 1997). This outlook is particularly interesting in our study considering the influence of the entrepreneur's profile (thus their history) on their application of a strategy for sustainable innovation. In addition, the evolutionary theory examines the importance of the setting of the innovative process, situating innovation in a company, but without isolating it from its environment. Among the literature's numerous descriptions of innovation, the two that are the most widely used are: its nature (distinguishing between product innovation, procedures innovation, commercial innovation and organizational innovation) (Barreyre, 1975); and the radicalism of the innovation (distinguishing between radical or broken-up innovation, incremental innovation and the adaptation of the existing innovation) (Damanpour, 1991).

Beyond these multiple forms of innovation, Schumpeter (1935) insisted on the importance of technological innovation. However, Carrier and

Garand (1996) show the difficulty of limiting the perimeter of technological innovation, denouncing an overuse of the word 'technological'. According to these authors, 'an innovation cannot be qualified as technological unless its major effect is to transform the study of technical knowledge affected by a domain or a particular sector' (Carrier and Garand, 1996: 14). The increasing power of sustainable development in political analysis contributes to the multiplication of the comments about the link between innovation and sustainable development (Mathieu, 2008). Numerous authors prefer the term 'environmental technology' when talking about technologies built around sustainable development. Authors such as Patris et al. (2001) distinguish additive environmental technologies (allowing for the reduction of toxic emissions) from integrated environmental technologies (aiming for a reduction at the source of the negative impact of products or procedures for the environment).

Concerning this analysis, innovation, a central theme of management, forms one of the major characteristics of entrepreneurship likely to generate performance in the domain of small businesses. This leads to the question of how we can combine this innovation with the challenges of sustainable development.

Strategies for Sustainable Innovation

In the statement put in place by the French state following the Grenelle de l'Environnement, the ministers responsible for industry and ecology began a process that aims to promote the development of eco-industries in France. Eco-activities produce goods and services used to measure, prevent, limit and rectify environmental impacts (IFEN, 2008).[1] The field of eco-business is restricted to establishments that ensure the sales of marketable goods and services. According to a 2008 study by the Boston Consulting Group,[2] the development of eco-industries is particularly significant in France, which already has a solid industrial base, estimated at €60 billion of activity per year and 400 000 jobs in 2008. Combining such an industrial policy for developing eco-industries with implementing the objectives of the Grenelle de l'Environnement could represent a direct increase of the activity in France of €50 billion per year and 280 000 new jobs by 2020. It would lead to an annual reduction of CO_2 emissions of 80Mt, a positive impact on the commercial scale representing €25 billion per year (in reduction of gas, petrol and raw material importation) and a growth in French buying power when investments in energy efficiency offer a rapid return on investment. 'Beyond eco-industries, it's the whole of the French economy that is affected by environmental and energy problems and the desire to become "green". It consists of the consumer's discernible expectation, faced with which the industry must organize itself and within which the industry can find a competitive edge' (BCG, 2008: 3).

Several papers in the area of sustainable development strategies can be found in the existing literature. Bellini (2003) compares the additive logic (where the enterprise does not question its decision-making process) to the systematic logic (where the integration of the environment deeply modifies the decision structure). Based on this, the author distinguishes between eco-defensive strategies (using an exclusively financial logic), eco-conformist strategies (which are happy to respect norms imposed by the regulations), and eco-sensitive (or proactive) strategies (going beyond the rules in effect). Bellini (2003) specifies that only the eco-sensitive strategies reveal the true nature of the enterprise. As for Martinet and Reynaud (2004), they identify three types of strategic attitudes:

1. A wait-and-see attitude on the leader's behalf, or in a more pragmatic way, to the lack of resources (human, financial).
2. An adaptive attitude aiming to conform to the legislation (DiMaggio and Powell, 1983) or to anticipate the market's future expectations for a 'responsible' offer;
3. A proactive attitude, that is the desire to play an innovative role within the specific sector of activity.

Mathieu (2008) suggests combining these sustainable development strategies with the environmental technologies deployed in certain organizations. She singles out an adaptive (or eco-defensive) approach characterized by incremental (or even marginal) environmental innovations of a proactive (or eco-sensitive) nature to be the only approach worthy of bringing the enterprise forward on the route to sustainable development (Hart, 1995). Proactive strategies are not often seen in business because they assume that organizations have at their disposal the appropriate management tools and significant financial means (Mathieu and Soparnot, 2007). Businesses that adopt a proactive strategy aim for long-term performance based on cost reduction, legitimacy and an asset in terms of differentiation strategy (for example brought about funding by government and regional authorities). 'For these businesses, the ecological/social data is strategic. Ecology and social issues are no longer considered to be decision-making criteria, but like economic issues, as outcomes of action' (Mathieu and Soparnot, 2007: 15). According to Hart (1995), only radically innovating options guarantee the long-term survival of an organization.

Technological innovation is the outcome of a collective process. The success of such a process is based not only on skills and financial resources, but also on the ability to introduce the innovation into the business's technical and socio-economic environment, which is what Latour (1989) qualifies as the 'recruitment of allies'. Currently, the most well-known initiatives for sustainable development come from big businesses. What small businesses do is often ignored (Schoenberger-Orgad and McKie, 2005). Often, these initiatives suffer from a lack of legitimacy for reason of their obscurity. Small businesses are however not absent from this movement (Observatoire des PME européennes, 2002), but remain timid when it comes to communicating about the subject.

Entrepreneur Profiles and Innovations

Studies of the influence of the entrepreneur's profile on various aspects of management have been the topic of several publications (Boissin et al., 2008). When it comes to attitude in regard to innovation, Miles and Snow (1978) identify four types of entrepreneurs, that is, defenders, prospectors, analyzers and reactors. In his study, Geindre (1998) suggested that prospectors and analyzers were predisposed to active networking. This answers their need to access the most up-to-date information about the evolution of their sector of activity.

On a larger scale, as early as the 1970s, authors show that the personal characteristics of small business leaders had an influence on the degree of social development. For Laufer (1975), an interaction exists between the entrepreneur's personality, the developmental conditions of his business, and his way of running the firm. Ever since, research has shown contradictory results. In 1994, Smith and Oakley showed the existence of a positive relationship between the age and ethics of small business leaders. On the other hand, according to Ede et al. (2000) age does have a significant effect on ethics in a small business, but in the opposite direction, bearing in mind that young entrepreneurs tend to react more ethically than their older colleagues. Ede et al. (2000) did not find a significant relationship between either gender or education and small business leaders' ethics. These results contradict those of Spence and Lozano (2000) who identified that education and training play an important part in small business leaders' ethics. In a study of 41 small American businesses published in 1997, Quinn highlighted the link between a leader's personal ethics and his attitudes when faced with ethical problems in his firm. Courrent (2003) defends the thesis that the ethic of an organization (in this case very small businesses) is easily compared to the leader's ethics. When it comes to gender, Alan and Alan (2001) conclude that women are significantly more ethical than men.

In order to analyze the relationship between the profile of the entrepreneur and his sustainable innovation strategy, we use the table suggested by Daval et al. (2002: 55) for which 'the objective is not so much creating

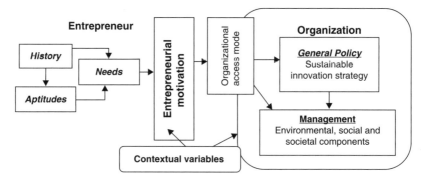

Source: Simplified model (inspired by Daval et al., 2002: 9).

Figure 9.1   Creator profile and sustainable innovation strategy

a meta-typology but to organize these elements into an integrative table. It's about proposing a tool that summarizes the diversity of entrepreneurs, without seeking generalization or determinism'. Figure 9.1 presents the double advantage of being built upon the existing models and of separating the analysis of the organization and the individual. The organization of the business is thus considered to be a response to the leader's expectation. The first three central categories treat the entrepreneur as an individual. We find in the first position his personal history (intrinsic characteristics, education, experience, relationship structure). This history conditions the second category, which is made up of his entrepreneurial aptitudes or his individual character traits (self-confidence, competitiveness, drive). These first two categories together determine the third, which is the nature of the entrepreneur's needs (security, autonomy, recognition, power, self-realization). 'The translation of the entrepreneur's need(s) into action assume that it (they) has (have) finally been expressed as a form of conscious motivation, then organizational actions' (Daval et al., 2002: 65). On an organizational level, two categories emerge from the authors' analysis: general policy and management (which grouped together the elements related to actions and the forming of a strategy). Among the elements of general policy, Daval et al. (2002) mention long-term survival and the creation of value. According to us, these two elements are omnipresent in sustainable innovation strategies, which makes this table useful. The contextual variables (environmental opportunities, socio-economic variables) become activating elements, conditioning and starting the entrepreneurial phenomenon in an evolutionary logic (Tremblay, 1997).

AN EXPLORATORY EMPIRIC STUDY: THE SMALL BUSINESS ECODAS

This section seeks to apply the model from the previous section. After having briefly introduced the firm ECODAS and having analyzed the firm's strategy, we can examine the influence of the entrepreneur's profile and the contextual variables on the sustainable innovation strategies.

Presentation of the Firm ECODAS

Before introducing the small business ECODAS, we should take another look at the methodology of this research. It is of a qualitative nature (Hlady Rispal, 2002) by reason of the character of the topic being studied (Wacheux, 1996). 'Putting in place the process of qualitative research above all wants to understand the why and the how of events in concrete situations' (Wacheux, 1996: 15). When it comes to an emerging theme, we have naturally highlighted a qualitative methodology centralized on a case study (Yin, 1994; Hlady-Rispal, 2002). According to Stake (2000: 345), 'Case studies have become one of the most common ways to do qualitative inquiry.' The analyzed unit for Yin (1994) is, in this case, the small business. The case study is based on treating secondary data (press articles, websites, testimonies given during regional professional events such as the Annual Day for Sustainable Development and Businesses 2008 in Lille or 'Diversity, An Advantage in Regards to the Crisis' from the conference in March 2009 in Marcq-en-Baroeul).

A global leader in its domain, ECODAS is a small business in the north of France specializing in the design and manufacture of medical waste treatment machines using shredding and sterilization technologies. Created in 2000 by J.S., the small business of 18 employees achieved sales figures of €4.6 million in 2010. The business exports 80 percent of its production to 60 countries, notably in North Africa, the Middle East, Russia and South America. Internationally, this small business collaborates with about 30 local hospital material distributors, the agents of foreign clients. ECODAS received the 2002 Trophy for International Promotion from the Secretary of State and External Commerce[3] and in 2008 the Eurasanté developers Club Trophy for Support for Innovation and International Development. In France, there are nearly 300 hospital establishments[4] which now resort to this efficient, ecological and affordable technology.

In spite of its modest size, the small business is developing a strategy for sustainable development based on a technological innovation. 'Innovation is processing without polluting and our product range responds to the needs of small hospitals as well as the bigger structures'.[5] J.S. has developed an innovative procedure that is entirely automatic, combining shredding and sterilization of the wastes produced by treatment activities at risk of infection[6] (DASRI) in a closed compact vessel. The wastes are introduced in the upper compartment of the machine fitted with a grinder, then transferred to an autoclave in the lower compartment where they are disinfected. The machine is water vapor-heated at 138°C under 3.8 bars of pressure.

After a totally automatic disinfection cycle lasting 20 to 40 minutes (according to volume), the ground-sterilized material can rejoin the recycling of regular refuse. The machines are made of recyclable stainless steel, and conform to the sustainable development strategy of the small business. The conception and development of this process makes ECODAS's leader a prospector according to Miles and Snow (1978). ECODAS's strategy for sustainable innovation is characteristic of proactive behavior (Martinet and Reynaud, 2004; Mathieu and Soparnot, 2007) and eco-sensitive (Bellini, 2003), as the methods address the environmental as much as the social component, a key element being the creation of value for this small business and its stakeholders.

Analysis of ECODAS's Strategy for Sustainable Innovation

From an environmental point of view, ECODAS's innovative process has numerous advantages:

- easy to use, as mainly automatic, simplified maintenance and little staff training needed (one day only);
- reliability thanks to shredding technology and sterilization in a closed and compact unique vessel, which eliminates all handling of waste;
- efficiency through a reduction of 80 percent of the waste volume;
- ecology thanks to the absence of smoke emission, chemical waste, radiation and to the elimination of waste at the earliest moment in the production process;
- economy in that the on-site installation in the hospitals reduces the transport of wastes and the treatment costs;
- the treatment process for hospital waste developed by ECODAS is economically profitable, as shown in Table 9.1.

ECODAS's clients believe the economic argument to be most important, a leading example being the regional center for the fight against cancer, the Léon Bérard in Lyon. It has used one of ECODAS's treatment machines since May 2006. As the center's logistics manager explains:

Table 9.1   Comparative study of different treatment processes for hospital waste

Technology	Microbiological Efficiency (level IV*)	Integrated pulverization	Capacity (kg/h)	Cost (Euro/kg)
ECODAS	Yes	Yes	30–300	0.05–0.09
Autoclave and pulverization	No	No	400–500	0.10–0.14
Radiation	No	Yes	200–300	0.14–0.18
Micro-wave	No	Yes	100–00	0.12–0.22
Pyrolysis/Plasma	No	No	300–500	0.14–0.34
Chemical	No	Yes	100–500	0.24–0.52
Incineration	Yes	No	Off site	0.64–1.04

Note: * bacillus Stearothermophilus, reduction by 10^7.

Source: ECODAS.

I won't deny that initially, the motivation was purely economic: our outside providers for waste treatment were increasing their rates in a manner that wasn't compatible with the increase in our budget: around 12 or 13 per cent per year on average! The second motivation was to decrease risk, we had had a couple of incidents of employees being pricked or cut. In addition, we had a desire to respect the environmental regulations: this process generates no discharge into the atmosphere or effluents . . . [In terms of cost, the results speak for themselves] . . . This procedure has allowed us to save 110 000 Euros annually on production costs for an initial investment of 370 000 Euros for the materials. (Mathis, 2007: 17)

On the economic level, the case study of ECODAS shows the strong link between economic performance and sustainable innovation strategy (Mathieu and Soparnot, 2007). If ECODAS's clients gain in adopting this process, the business itself also benefits from enviable economic results, as illustrated in Table 9.2. Its sales figures increased over 46 percent between 2005 and 2008.

Lastly, the social component is not forgotten by this northern small business, proving that it is possible to reconcile all three pillars of sustainable development. The business is characterized by a real policy for cultural diversity.[8] True to his convictions, ECODAS's chief signed the Charter for Diversity on June 14, 2005, joining 84 businesses in Nord Pas-de-Calais in this initiative (figures from July 2009). Launched in January 2004, the Charter for Diversity is a commitment to condemn employment discrimination and opt to act in favor of diversity undertaken by each business that signs, regardless of its size. As J.S. explains: 'It is not volition, it's

Table 9.2   ECODAS's key figures since 2004

Year	Sales figures	Result	Employees
2010	€4.6 million	722 689	18
2009	€5.6 million	549 745	21
2008	€7.5 million	1 461 000	22
2007	€5.6 million	720 846	18
2006	€3.9 million	404 470	15
2005	€5.2 million	797 395	15
2004	€4.4 million	841 551	17

Source: Infogreffe.

an opening of the mind. Each time that we have hired, we have taken the best candidates. Result: half of our employees are immigrants.' The key phrase here is 'equality of chances'. Since June 2008, ECODAS has been a member of the Pacte Mondial (Worldwide Pact), and more specifically of the Principle no. 6, which is eliminating discrimination in employment and in the workplace. This formal commitment to integrate the principles of the Pacte Mondial reflects the management and human resources policies of this small business. In 2008, 29 percent of the employees were of foreign nationality and the ratio of *cadres* (upper management) of foreign nationality was 50 percent. As for the percentages in terms of employee age by range, 9 percent were less than 25 years old, 43 percent 25–35 years old, 29 percent 35–55 years old, and 19 percent aged over 55. Strengthened by this human and cultural diversity, ECODAS looks to involve its employees in various humanitarian actions through associations such as BAYTI[8] (help for children in difficulty in Morocco) and AMCV[9] (surgical operations for underprivileged children in Morocco).

To sum up, the case study of ECODAS illustrates that businesses are able to make technological choices that can be economically advantageous for the clients, ecologically sustainable and socially responsible. In the next subsection, we analyze the influence of the entrepreneur's profile on this sustainable strategy.

The Influence of the Entrepreneur's Profile and Contextual Variables on Sustainable Innovation Strategy

An automation engineer with a degree in management from the IAE[10] in Lille, J.S. started working on his hospital waste treatment project in 1993. After a long journey and a lot of persuasion, the chief of ECODAS came to have his process validated and create his business in 2000. One of

J.S.'s dominant character traits is without a doubt his tenacity, because the gamble of modifying the legislation was a battle not easily won. In the early 1990s, J.S. was at the head of a small business that fabricated autoclaves for the textile industry. Faced with the crisis in this sector, the entrepreneur made the choice to change to a different market. Strengthened by 20 years of industrial know-how, he decided to make skills in thermal process management more profitable in a new sector of activity. At this time, hospital wastes were being burned, as per the legislation in effect. The hospitals had their own incinerators or entrusted their waste to approved treatment centers. However, incinerating hospital wastes was not without consequences on the environment. This is where J.S. had the idea to propose to hospitals a more ecological, economical alternative for treating their wastes. The path of an entrepreneur is littered with obstacles, beginning with the regulation that authorizes only incineration. Also, J.S. spent over a year convincing public authorities of the benefit of his production process. Stubbornness paid off; health and environmental ministers accorded him a six-month exemption, subject to having the process validated in the hospital setting. The hospital center in Roubaix agreed to carry out a full test program. Thanks to the financial support of ANVAR,[11] a first machine was conceived and fabricated. This project lasted 18 months. ECODAS's process was approved by the Superior Board of Directors for Public Hygiene in France in 1994. In 1998, J.S.'s previous business filed for bankruptcy. From the moment that ECODAS was created in 2000, he rehired half of his previous business's employees and purchased the property in order to create his new business.

ECODAS's economic performance is based on an original economic model that is based on disruptive innovation technology, associated with proactive management for a small business. The managerial choices made by J.S. reflect the education that he has had, both technical and administrative. His education in management at the Institut d'Administration des Entreprises (IAE) in Lille allowed him to better understand the risk factors associated with the deployment of tools which are often perceived to be the prerogative of large businesses, such as intellectual rights, business intelligence, the formalization of human resource management policies (Worldwide Pact for Diversity), or the international development of emergent countries and markets.

When it comes to intellectual rights, ECODAS's effort is exemplary for small businesses of this size. The worldwide organization for intellectual property published a report in 2008, giving global statistics about patents, the World Patent Report 2008. This report highlights the large number of patents for environmental technologies. In fact, between 2001 and 2005, 1123 patents were declared for this domain. France is ranked in fourth place worldwide for the declaration of patents in the environmental technology domain, yet lies far behind the top three leading countries: Germany, the United States and Japan (IFEN, 2008). ECODAS's process was approved according to American (ASME), Chinese (MLSE), European (CE), Japanese (MHLW Japan), Polish (UC Poland) and Russian (Gosstandart) norms in addition to ISO 9001 certification.

J.S. speaks English, Arabic and Portuguese, which allows him to actively participate in the international development of his small business. On average, he travels abroad at least once a month, either for professional fairs or for developing his network of distributors. Since the beginning, the business has exhibited at several professional fairs such as Medica (Dusseldorf), Hôpital Expo (Paris) and Pollutec (Paris). Even more surprisingly, the small business has developed economic and strategic intelligence awareness actions, piloted by the marketing manager (who happens to be J.S.'s wife): 'The goal is to be on the look-out for all of the information that could possibly be able to help the business develop in France and abroad and to allow those who need to benefit from this to do so' (Vernier, 2004: 37).

The precocious development of ECODAS internationally is remarkable for a business of its small size. To give an example, at the end of 2003, a delegation led by the Chinese minister of the environment took place at the firm at the initiative of ADEME.[12] In November 2004, J.S. participated in the Pollutec China fair in Shanghai in order to promote French know-how in the matter of waste treatment. Since June 2009, ECODAS is officially one of the first European eco-enterprises (eco-businesses),[13] meaning a business that produces goods or services capable of preventing and correcting damage to the environment, such as air, water and land pollution, as well as the problems linked to waste materials and noise. This label is recognition of the small business's commitment through the years but above all allowed it to take advantage of foreign markets (particularly in Northern Europe), which were essential for the business. In 2008, this small business opened a representative office in the United States, obtaining certification in about 20 states.

Of a humanistic nature, at 44 years of age, ECODAS's chief leader has not forgotten his Moroccan origins. J.S. explains that he has never personally suffered from integration problems. In 1997, he founded an association aiming to assist immigrant youth in their economic development (Synergie). J.S.'s ambition is to diffuse the 'Big Brother' culture by sharing experiences. The goal is to help young immigrants find their place in the world of work. As a sign of his optimism, J.S. explains: 'I hope that in a few years, Synergie won't have to exist. We'll have a multicultural model.'

He refuses to use the term 'positive discrimination'. For him, it's about hiring the best without prejudices.

In addition, J.S.'s commitment to the local business networks is significant: he is a member of the Comité Grand Lille (Greater Lille Committee) at the head of a group geared toward integration, as well as a member of the CCI (Chamber of Commerce and Industry) of the Lille metropolitan area. ECODAS is a member of the International ADEME club, the objective of which is to support French eco-enterprises (eco-businesses) in their growth within foreign markets. This intense involvement of ECODAS's leader reminds us of Geindre's (1998) proposition when it comes to the propensity of prospectors to practice active networking.

Beyond the profile of an entrepreneur, the role of contextual variables on the adoption of sustainable innovation strategies is far from negligible in the context of small businesses, confirming the pertinence of Daval et al.'s (2002) model. Environmental regulation is the primary factor for the long-term growth of eco-enterprises (eco-businesses). Historically, French environmental regulation (and that of the European Union) primarily targeted water and waste domains (IFEN, 2008). It thus favored the businesses which specialized in these areas. The evolution of a regulatory framework is particularly favorable for the innovation developed by ECODAS. The European Waste Directive 2008/98/CE published in the *Official Journal of the European Union*, L312 November 22, 2008, constitutes the new reference text for European waste management policies. It establishes the recycling goals to be reached by 2020: 50 percent of the weight of household waste must be recycled. In fact, the sales perspectives for the years to come are significant in the domestic market because: 'If we are technological precursors, then in practice, we're behind. In France, 80 per cent of hospital waste is still burned . . . Globally, we are far behind the countries of northern Europe. The European Waste Directive asks that each country reach the goal of recycling 50 per cent of its waste by 2020. France is at less than 30 per cent for its household waste.'[14] One of ECODAS's strengths seems to be its ability to envisage the environment in another way, and see it as an opportunity for innovating. The innovation policy of this small northern business involves a strictly technological and organizational framework. It questions the small business's 'traditional' model ever-present in current literature about entrepreneurship.

CONCLUSION

The purpose of this chapter was to gain a better understanding of why economic representatives make the choice of a sustainable innovation strategy, by showing the link between the entrepreneur's profile and the development of such a strategy. To do this, a case study was produced in order to explore the reality of a sustainable development strategy based on a technological innovation introduced at ECODAS, a small business in the north of France. There are three parts to this research:

1. The analysis of ECODAS shows the realization of a sustainable innovation strategy reconciling economic success, social commitment and respect for the environment (in agreement with the recent political and social developments).
2. The chapter points out the influence of the entrepreneur's profile on the adoption and implementation of a sustainable innovation strategy. The director of ECODAS is typically innovative in terms of Miles and Snow's definition (1978).
3. Particular stress was placed on the role of contextual variables influencing the adoption of a sustainable innovation strategy. This refers back to the work on the innovative environment within the framework of evolutionary theory (Dosi, 1988; Nelson and Winter, 1982: Le Bas, 1995).

To summarize, the key factors for the success of a sustainable innovation strategy for small business are based on the elements of the entrepreneur's profile (social capital, audacity, tenacity) associated with the correct environment for the putting in place of its strategy, particularly in terms of the law regarding sustainable development. Certainly, the results only suggest tendencies that would have to be further explored in research, as we have only studied one unique business.

From a managerial point of view, it seems important to help entrepreneurs become aware of the movement towards a green economy. This implies educating the organizations, giving assistance to small business leaders regarding sustainable development and, notably, encouraging them to integrate it more often into their business plans. Initiatives (marginal in the French context) can be highlighted such as the Réseau Entreprendre® (Entrepreneurial Network), which since 2009, through assistance to entrepreneurs towards the growth of their businesses, has proposed that as soon as they start preparing their project, they take into consideration the components of sustainable development just as they would the other elements such as the international dimension and encouraging innovation. As Welsh and Herremans wrote (1998: 154): 'in the short term, this process might be elaborate, time consuming, and unnecessary, but in the long term, it ensures survival of the company'. Moreover, the conditions for the profitability of businesses are rapidly evolving due to an ever-changing regulatory and social framework. Actions considered non-profitable in the current economic context will undoubtedly be profitable in the years to come.

NOTES

1. See IFEN (2008).
2. See BCG (2008).
3. The MOCI n°1635 January 29, 2004: 'ECODAS – A global leader resulting from a successful restructuring', by Geneviève Hermann.
4. In 2005, barely 100 machines were installed in France and in foreign countries (mostly in the Maghreb).
5. See Ducuing (2006).
6. These wastes can contain syringes, needles, plastic, glass and other non-fabric materials.
7. BAYTI was founded in 1994 and works in the domain of the familial reintegration, reintroducing children in difficult situations (homeless, forced labor, victims of violence, abandoned and delinquent children) into a scholastic and socio-professional system.
8. *La Voix du Nord* (Northern Voice) September 26, 2007, 'Social diversity: ECODAS even recycles prejudice', by Thibaud Vuitton.
9. AMCV is the Moroccan association for Visceral Surgery.
10. IAE: Institute of Business Administration.
11. ANVAR: National Agency for the Validation of Research.
12. ADEME: Agency for the Environment and the mastering of Energy.
13. Label given by AFNOR (French Association for Norms).
14. *La Tribune*, May 15, 2009, 'With Ecodas waste takes the heat'.

REFERENCES

Abrassart, C. and F. Aggeri (2007), 'Quelles capacités dynamiques pour les stratégies de développement durable des entreprises? Le cas du management de l'éco-conception', *Actes de la XVIème Conférence de l'AIMS*, Montréal.

Alan, K.M.A. and C.B.T. Alan (2001), 'Marketing ethics and behavioral predisposition of Chinese managers of SMEs in Hong Kong: Global perspective', *Journal of Small Business Management*, **39**(3): 272–8.

Barney, J. (1991), 'Firm resources and sustained competitive advantage', *Journal of Management*, **17**: 99–120.

Barreyre, P.Y. (1975), 'Radiographie de l'innovation', in *Encyclopédie du Management*, Vol. 2, Paris: Klumer.

BCG (2008), 'Eco Tech 2012 – Strategic Plan Assessment', 2 December.

Bellini, B. (2003), 'Un nouvel enjeu pour l'entreprise: la prise en compte de la protection de l'environnement dans son management – Etat des lieux et perspectives', Atelier Développement Durable de l'AIMS, Angers, 15 May.

Boissin, J.P., M.C. Chalus-Sauvannet, B. Deschamps and S. Geindre (2008), 'Profils de dirigeant et croissance des jeunes entreprises', *Actes de la Conférence de l'AIMS*, Nice.

Ducuing, O. (2006), 'Ecodas soigne les déchets à risques des hôpitaux', *Les Echos*, **19815**, 14 December.

Camison-Zomosa, C., R. Lapiedra-Alcani, M. Segarra-Ciprés and M. Boronat-Navarro (2004), 'A meta-analysis of innovation and organizational size', *Organization Studies*, **25**(3): 331–61.

Carrier, C. and D. Garand (1996), 'Le concept d'innovation: débats et ambiguïtés', *Actes de la Vème Conférence de l'AIMS*, Lille.

Courrent, J.M. (2003), 'Ethique et petite entreprise', *Revue Française de Gestion*, **29**(144): 139–52.

Damanpour, F. (1991), 'Organizational innovation: a meta-analysis of effects of determinants and moderators', *Academy of Management Journal*, **34**(3): 555–90.

Daval, H., B. Deschamps and S. Geindre (2002), 'Proposition d'une grille de lecture des profils d'entrepreneurs', *Revue Sciences de Gestion*, **32**: 53–74.

Davis, J. (1992), 'Ethics and Environmental Marketing', *Journal of Business Ethics*, **11**(2): 81–7.

DiMaggio, P.J. and W. Powell (1983), 'The iron cage revisited: institutional isomorphism and collective rationality in organizational fields', *American Sociological Review*, **48**(2): 147–60.

Dosi, G. (1988), 'Sources, Procedures, and Microeconomic Effects of Innovation', *Journal of Economic Literature*, **26**(3): 1120–1171.

Ede, F.O., B. Panigraphi, J. Stuart and S. Calcich (2000), 'Ethics in small minority business', *Journal of Business Ethics*, **26**: 133–46.

French Institute for the Environment (IFEN) (2008), 'Croissance soutenue de l'activité des éco-entreprises entre 2004 et 2007', *Le 4 pages*, **127** (November).

Geindre, S. (1998), 'Profil de dirigeants et réseautage en PME: proposition d'un cadre de recherche', *Actes de la 7ème Conférence de l'AIMS*, Louvain-la-Neuve.

Hart, S. (1995), 'A natural resource based view of the firm', *Academy of Management Review*, **20**: 986–1014.

Hlady-Rispal, M. (2002), *La méthode des cas – Application à la recherche en gestion*, Bruxelles: De Boeck Université.

Jaouen, A. (2008), 'Le dirigeant de très petite entreprise: éléments typologiques', *Actes du Congrès International Francophone sur l'Entrepreneuriat et la PME*, Louvain la Neuve, 29–31 October.

Julien, P.A. and R. Lachance (2006), 'Colombo, Holmes, Maigret, de Baskerville et l'entrepreneuriat régional', in C. Fourcade, G. Paché, R. Perez (eds), *La stratégie dans tous ses états*, Caen: Editions EMS, pp. 323–38.

Latour, B. (1989), *La science en action*, Paris: La Découverte.

Laufer, J. (1975), 'Comment devient-on entrepreneur?' *Revue Française de Gestion*, pp. 11–26.

Le Bas, C. (1995), *Economie de l'innovation*, Paris: PUF.

Martinet, A.C. and E. Reynaud (2004), *Stratégies d'entreprise et écologie*, Paris: Economica.

Mathieu, A. (2008), 'Eco-stratégies et réponse industrielle aux attentes sociétales: quelles implications?' *Actes du 3ème Atelier DD de l'AIMS*, Lyon.

Mathieu, A. and E. Reynaud (2005), 'Les bénéfices de la responsabilité sociale de l'entreprise pour les PME: entre réduction des coûts et légitimité', *Revue de l'économie méridionale*, **53**(211): 357–80.

Mathieu, A. and R. Soparnot (2007), 'L'appropriation du concept de développement durable en entreprise: un générateur d'innovation', *Actes de la XVIème Conférence de l'AIMS*, Montréal.

Mathis, D. (2007), 'Déchets: une experience, une stratégie', *DH magazine*, **111**(February–March): 17.

McWilliams, A., D.S. Siegel and P.M. Wright (2006), 'Corporate social responsibility: strategic implications', *Journal of Management Studies*, **43**(1): 1–18.

Miles, R. and C.C. Snow (1978), *Organizational Strategy, Structure and Process*, New York: McGraw Hill.

Nelson, R.R. and S.G Winter (1982), *An Evolutionary Theory of Economic Change*, Cambridge, MA: Harvard University Press.

Observatoire des PME européennes (2002), 'Les PME européennes et les responsabilités sociale et environnementale', rapport n°4.

ECD (2006), *Etude économique du Canada*.

Patris, C., G. Valenduc and F. Warrant (2001), 'L'innovation technologique au service du développement durable', Synthèse du rapport final du programme 'Leviers du développement durable', Bruxelles.

Schoenberger-Orgad, M. and D. McKie (2005), 'Sustaining Edges: CSR, postmodern play and SMEs', *Public Relations Review*, **31**(4): 578–83.

Schumpeter, J. (1935), *Théorie de l'évolution économique*, Paris: Dalloz.

Soparnot, R. and E. Stevens (2007), *Management de l'innovation*, Paris: Dunod.

Spence, L.J. and J.F. Lozano (2000), 'Communicating about ethics with small firms: experiences from the UK and Spain', *Journal of Business Ethics*, **27**: 43–53.

Stake, R.E. (2000), 'Case Studies' in Denzin N.K. and Y.S. Lincoln (eds), *Handbook of Qualitative Research*, London, UK and Thousand Oaks, CA, USA: Sage.

Tremblay, D.G. (1997), 'Innovation, management stratégique et économie: comment la théorie économique rend-elle compte de l'innovation dans l'entreprise?' *Actes de la VIème Conférence de l'AIMS*, Montréal.

Van de Ven, A.H. (1986), 'Central problems in the management of innovation', *Management Science*, **32**: 590–607.

Vernier, S. (2004), 'Ecodas, hospital waste under pressure', *FACE*, **161**: 36–37.

Verstraete, T. and A. Fayolle (2005), 'Paradigmes et entrepreneuriat', *La Revue de l'Entrepreneuriat*, **4**(1): 33–52.

Wacheux, F. (1996), *Méthodes qualitatives et recherche en gestion*, Paris: Economica.

Welsh, C.N. and I.M. Herremans (1998), 'Treadsoftly: adopting environmental management in the start-up phase', *Journal of Organizational Change Management*, **11**(2): 145–56.

Wernerfelt, B. (1984), 'A resource-based view of the firm', *Strategic Management Journal*, **5**: 171–80.

Yin, R.K. (1994), *Case Study Research – Design and Methods*, London: Sage Publications.

10. Benchmarking sustainable construction technology in the building and transportation sectors

Salwa Beheiry and Ghassan Abu-Lebdeh

INTRODUCTION AND BACKGROUND

This chapter reviews the principles of benchmarking sustainable construction (SC) technology and techniques in the building and transportation sector. For the purposes of this chapter, building construction includes residential and commercial property, whereas transportation projects include facilities such as roadway, bikeway and walkway sections; junctions including roundabouts, at-grade signalized and non-signalized intersections, and grade-separated intersections; bridges including piers, girders and decks; tunnels; runways, taxiways, and holding areas; and guideways including railways; bus stops, rail and metro stations and transfer points; pavements (i.e. road surfaces), both rigid and flexible, and base and sub-base layers; and drainage facilities including open and closed channels, culverts and collectors. The chapter also provides a brief highlight of a benchmarking case study, using a pilot data sample and a follow-up data sample of residential and commercial buildings.

It has been asserted over the past two decades that benchmarking is a crucial component of a mature project delivery system. It enables organizations to gather information and to understand project performance and best practices. Whether the driver is cost, schedule, quality or all of these, benchmarking is an essential part of any continuous improvement process (Hwang et al., 2008). Process benchmarking is the most recent development. It offers additional benefits over product or performance benchmarking by enabling work to be viewed as a series of holistic transformation events with identifiable inputs and outputs. Moreover, benchmarking could be a powerful tool in investigating and managing change on construction projects (Garnett and Pickrell, 2000).

THE BUILDING SECTOR

A study by Kibert (2008) emphasizes the construction industry's significant impact on our planet's resources and environment. The built environment has direct and multifaceted impacts that also tend to be long-lasting. In the United States, building materials and components production, in addition to the construction process, require the extraction of approximately 6 billion tons of basic materials annually. In other words, the construction industry, which is about 8 percent of United States (US) gross domestic product (GDP), uses up to 40 percent of extracted materials in the US.

Kibert indicates that some estimates suggest as much as 90 percent of all materials ever extracted in the United States are for buildings and infrastructure. Construction waste in the United States, on the other hand, is generated at the rate of about 0.5 tons per person annually. This is approximately 5–10 lbs per square foot of new construction and 70–100 lbs per square foot of renovation work. These figures are exacerbated by the limited levels of recycling and reuse of demolition debris and waste from construction activities.

Hill and Bowen (1997) suggest that the term 'sustainable construction' was originally proposed to describe the responsibility of the construction industry in attaining sustainability. They go further to categorize the principles of sustainable construction into four pillars; social, economic, biophysical and technical. On the other hand, the United Nations Environment Programme (UNEP) Agenda 21 states that SC technologies enhance the quality of the construction process and its products by reducing resource use.

A 2003 study by Kibert indicates that substantial progress has been made towards the adoption of sustainable construction maxims. A green building movement has emerged in many countries. The manufacturing and services sectors have started creating new 'green tagged' products. Numerous research centers are appearing in many educational and commercial organizations investigating sustainable construction strategies, technologies, materials and systems. National and local governments are adopting high-performance building standards. Educational institutions are offering undergraduate and graduate degrees in sustainable construction and related disciplines.

Nonetheless, Kibert (2003) points to the fact that until buildings and their supporting resource systems are totally based on renewable resources, are deconstructable and the components able to be totally reused and recycled, are integrated into a sustainable urban plan, are healthy for their occupants, process their waste using biological systems, and have a host of other similar features, sustainability in the built environment will be just a distant objective.

It can be clearly concluded that the universal movement towards sustainable construction, while real and robust, requires a substantial time frame to gain momentum in all parts of the world and produce fundamentally sustainable buildings, when compared to the current practices.

Energy consumption is the major environmental culprit as far as the built environment is concerned (Horsley et al., 2003). For instance, residential construction was nearly 4.2 percent of the US GDP in 2000, and residences consumed nearly 20 percent of total US energy consumption (Ochoa et al., 2005). Sustainable construction (SC) improves the environment by improving air and water quality, and reducing energy and water consumption, and waste generation. Sustainable technologies and techniques include such things as the use of composite materials, the use of recycled content in new products, new prefabricated housing systems, low energy use, water conservation and pollution control devices.

Acquaye et al. (2009) argue that sustainable engineering design and energy efficiency programs for particular construction subsectors can support the reduction of the overall energy use in the building sector. The study suggests that embodied energy analysis can be used as a construction design assessment tool in a sustainability matrix for building projects. The study is reserved on the straightforwardness of implementation because of data measurement inaccuracies. In the Acquaye et al. study, comparison between the data in a deterministic embodied energy (EE) model of a building and the stochastic EE model of the same building carried out using Monte Carlo simulation demonstrated a considerable variation in outcomes. This is mainly due to the lack of standardized specification of the energy intensity of building materials within industry requirements. The chosen material can sway the modeling data and reduce or raise the EE of buildings. More standardization will facilitate the effectiveness of the sustainable construction design assessment tool.

Several UN reports have stressed the need for lightweight and lower-energy-based materials, new assembly and disassembly techniques, and flexible building design (Beheiry, 2011). Flexible building design is based on a 'base' building concept. Buildings are designed to be durable but without specific interior fit-outs. This facilitates their future adaption for different user requirements over the building's lifespan. These reports also stated that construction materials technology has changed in the last decade, including reuse and recycling of materials and new engineered materials. Additionally, automation and off-site fabrication of components has increased, leading to alternative contracting strategies and newer organizational environments counterbalancing the sustainable technologies expenditure.

A specific example of SC technology could be the Terra Block Fabricator, investigated by Mehta and Bridwell (2005). The fabricator had the advantage of using local soil and labor to create high-quality building blocks. The study found that creative technology and the reliance on local natural resources and labor skills instead of imported construction materials increases the potential affordability of decent housing for the working and middle classes.

Another example is solar technology in buildings. For a very long time solar panels were not considered aesthetically pleasing in buildings, especially residential ones. This, coupled with their higher initial costs and lower efficiency compared to hydropower or fossil fuels, led to procrastination in solar and renewable energy research. Dunay et al. (2006) reviewed several refined techniques to incorporate solar technology in many facets of building projects without compromising aesthetics and finances.

A research study by Qian (2008) endorsed the notion that sustainable technology needs to be measured and assessed to be promoted. The relevance of green building assessment and measurement tools is now widely accepted in promoting sustainable construction. Green building assessment systems are emerging as the way forward in future building technologies. However, green building assessment tools range from the international to the local, and different organizations choose a combination of both.

Qian stresses that the value of an assessment tool is rated based on its potential application in the relevant industries. That is, it must be conceived with reference to the native economic requirements. It must also observe the regional indicators in order to create the assessment system, the implementation tools and strategies. Moreover, the local bylaws, environmental regulations and cultural values should be observed.

Qian (2008) introduced a set of high-level strategies for implementing green building assessment. These strategies include government policy to promote green building assessment, and standardizing green building assessment in accordance with the international norms, while customizing aspects of it to the local norms. Qian also suggests the establishment of a fundamental database of the assessment data, and training a critical mass of professionals in green building assessment.

Another study by Steinert (2008) stressed the increased popularity of sustainable construction in the last decade, especially in the developed world. The study indicated many positive impacts of sustainable construction such as improved efficiency in resource consumption and reduced impact on the environment. Other benefits include longer project life spans and measurable savings in life-cycle costs. Moreover, quality of life is improved by better indoor air quality and use of natural light. In most parts of the world personal, institutional and governmental initiatives have been started to promote sustainable construction, and standards have been created with which to measure sustainability.

On the other hand, developing countries are still mostly in the construction phase of their major basic infrastructure. The depletion of resources in these countries could continue for generations to come. Sustainable construction practices can be studied and assessed to be tailored to developing countries in order to avoid many of the problems that the developed world has experienced.

Steinert concludes that to focus on sustainable construction, a sustainable construction practices knowledge base should be generated to support government decisions, industrial organizations and educational institutions. The current status of the role of sustainability in developing and third world projects needs to be examined to build a knowledge base. Areas of improvement can then be identified and worked upon.

A study by Hazelton and Clements (2009) suggested that an analysis of resource waste triggered a synchronized, comprehensive reuse, recycling and waste management program. This effort reduced the off-site and on-site landfill requirements, the transportation impact and resource use. For instance, Bitou bush was reused from coastal protection zones to golf courses to recover green growth. Elements such as leachate in the groundwater from a landfill was diverted away from an existing lake and reused in golf course irrigation. The local sewer treatment plant was upgraded and an existing public pipeline was redirected and used to meet irrigation demands. Thus, the analysis contributed to the sustainability aspects of the project via addressing the existing environmental risks.

Khasreen et al. (2009) put forward life-cycle assessment (LCA) as an environmental impact estimation tool. LCA methodologies are highly applicable within the building sector. Khasreen analyzed LCA from a buildings perspective and examined its importance as a decision support system or tool. The research also reviewed life-cycle studies in the building sector within the last 15 years in Europe and the United States. The study highlights the lack of a universal and parallel database and assessment methodology to apply LCA within the building industry. The study also identified several areas for future research in environmental assessment.

On the other hand, a report by Augenbroe and Pearce (2012) attests that creating a more sustainable future is a long and ongoing process. Moreover, the short-term results are not always visible. The authors go on to suggest that for tools, policies and new technologies in the construction industry to work, they need to incorporate the relevant environmental policies.

A study by Yigitcanlar and Dur (2010) introduced a new urban sustainability assessment model entitled the Sustainable Infrastructure, Land-use, Environment and Transport Model (SILENT). The SILENT model is a geographic information system (GIS) application for urban sustainability indexing. The study promotes GIS applications as tools that can improve cooperation in decision-making among strategists and planners bent on promoting sustainable development.

A Korkmaz and Singh (2011) study asserts that good team communication is critical in sustainable projects since they require high interdependency among different building trades. For instance to minimize energy loss and maximize daylight, the electrical, energy and lighting trades have to communicate clearly. Their study examined team integration in a classroom project setting. The study confirms the role of team mechanics and dynamics, especially when working on a class-assigned sustainable project. The results show that more team communication can lead to better outputs, especially when team members also share common values, and rely upon and trust each other. The role of leadership in breeding better team communication even without team members having previous experience of working together was also observed.

A study by Dewlaney et al. (2011) quantified the increase or decrease in fundamental safety risk elements stemming from the design criteria and construction methods considered to achieve specific Leadership in Energy and Environmental Design (LEED) credits. In this study, fundamental risk is defined as the typical safety situations that exist with customary designs and construction processes and methods. The study highlighted that it is essential for designers and constructors to recognize and manage the increased safety risks linked with the more safety-conscious sustainable design and construction. In this study the influence of sustainable design and construction methods was quantified as direct multipliers (positive or negative) against the base situation. For instance, a 20 percent decrease in incidents associated with a LEED-triggered construction method would be defined as 20 percent lower than a conventional construction method. The Dewlaney et al. study stipulates that the knowledge created from the research would contribute to identifying the highest-risk design elements and construction activities and contribute to prioritizing safety resources allocation in response to high-risk elements.

Another technology that can assist in promoting sustainable construction strategies and practices is building information modeling (BIM). Bynum et al. (2012) argue that the use of BIM improves project planning and scheduling techniques. After a learning curve on BIM technologies, project cost is more controllable and project contingencies are minimized.

The Bynum et al. study examined the respondents' perception of the applicability of BIM for sustainable design and construction among designers and constructors. The questionnaire identified existing BIM technology and whether it can be used in sustainable design and construction. The results showed that 89 percent of respondents used BIM and 63 percent agreed that sustainable design and construction aspects were of importance in their corporation. However, most respondents indicated that sustainability is not a primary application of BIM and should be used predominantly for project coordination and visualization.

Bynum et al.'s study shows that most of the respondents perceive design–build and integrated project delivery (IPD) as the project delivery methods to use with BIM and to involve sustainability. Also 91 percent of the survey respondents believed that the schematic design phase is the optimal phase to implement BIM in sustainable design and construction. The study concludes that BIM is a recent technology but that as it expands in use, it will become a tool for sustainable design and construction.

Finally, Boyko et al. (2012) explore the use of technology to bring to life scenarios as a future visualization tool. The scenarios focus on the impact of uncertain futures on the performance of different indicators for sustainability in an urban regeneration context within a systematically quantified cadre. Boyko et al.'s paper introduced a toolkit through which the susceptibility or flexibility of a range of sustainability solutions can be tested systematically against a number of different future scenarios or outcomes. Different sustainable development or sustainable planning strategies are also tested against plausible outcomes. The aim of Boyko et al.'s research was to use technology to produce scenarios to visualize future outcomes from sets of sustainable practices and policies.

THE TRANSPORTATION SECTOR

Until recently, sustainable or green transportation implied human-powered or low-energy modes and/or vehicles using renewable and alternative fuels. In the current broader sense, however, there is far more to sustainability in transportation. But first an important difference is noted between a general building construction project and a transportation construction project: while a typical building construction project has planning, design, construction and occupation phases, a transportation project is 'operated' as opposed to 'occupied'. Operating can be viewed as the 'software' used to efficiently run and, thus, obtain the intended utility of the transportation project. For example, a traffic signal project goes through the planning, design and construction phases as do most other construction projects. The signal, or group of signals, can be operated (i.e., timed and varied based on spatio-temporal factors) to support different aspects of sustainability, such as countering congestion, improving safety, and so on. In addition, the signal hardware can be selected to be an energy-saving type (e.g., use of LEDs versus incandescent light bulbs). Because of the complex, large-scale, integrated and open nature (i.e., to other urban systems) of most transportation infrastructure projects, adoption of sustainability in planning, design, construction and operations is eminently significant.

To measure the use of technologies that promote sustainable transportation, it is necessary to identify the impacts that make a transportation project unsustainable. These impacts are: non-renewable fuel depletion; greenhouse gas emissions and the subsequent increase in global average temperature; global climate change in the form of flooding tunnels, coastal highways, runways and railways, and buckling of highways and railroads; poor local air quality; fatalities and injuries; congestion; noise pollution; low mobility as demonstrated by limited mobility for the poor, children, the elderly and the disabled; and ecosystem damage and disruption of natural habitats (TRB, 2004).

While some of those impacts can surface – perhaps in different forms or varying levels – at different stages of a transportation project, a close examination shows that some of those negative impacts are better addressed at one stage of a project (e.g., planning) than others (e.g., operating). The technologies necessary to address those impacts vary widely. More importantly, for such technologies to bear fruit, supportive policies must be in place.

This chapter addresses only sustainability on the construction side of transportation projects, although it can be argued that all aspects of a transportation project are interrelated and hence while the emphasis is construction, the presentation is relevant to all remaining aspects and phases of a transportation project. Technologies are discussed which directly or indirectly support eliminating or minimizing the above-mentioned negative anti-sustainability impacts. The technologies are divided into two broad areas: materials and practices. The presentation does not necessarily follow the order of the negative impacts noted above.

Use of renewable or recycled materials is common, especially in roadway and airport runway surfaces (Richardson and Jordan, 1994; Toutanji, 1996; Hoa et al., 2011; Walubita et al., 2011). Recycled material and special additives can be used in different types of surfaces, both flexible and rigid, with the benefits of prolonging the useful life of surfaces and minimizing the need for fresh component materials, and in some cases minimizing waste from old surfaces. Supportive policies and legislations, necessary to accelerate and formalize the use of recycled material in transportation construction, have been enacted and written into laws as early as the Federal Intermodal Surface Transportation Efficiency Act of 1991 (ISTEA, 1991). And in 1994, it was mandated that 5 percent of asphalt laid using US federal aid must contain recycled rubber from tires, and the law required the percentage to increase over time. Legislation with similar aims was considered in 44 states (Pennisi, 1992). Among other benefits, including energy savings, use of recycled material has significant positive impacts on the ecosystem.

Low-energy-consuming materials or components are now available as viable alternatives. Although in some cases initial costs are higher, use of such alternatives translates into significant savings over the life of a project. One such example is the use of LEDs versus incandescent bulbs in traffic signals (Evans, 1997). A LED traffic light uses 15 percent less energy. Given the extensive use of different forms of lighting in transportation projects, deliberate use of energy-saving alternatives can produce significant energy savings. Another interesting technology aimed at reducing energy consumption in transportation construction is the use of new hot mix designs that lower both the production and placement temperatures of hot mix asphalt (Prowell, 2007). Among the leaders in this area, COLAS's 3E Warm Mix is 40° to 45° C lower than conventional hot mixes (COLAS Group, n.d.).

Adopting and then constructing energy-efficient entire modes of transport is at the heart of sustainability. It is been shown that high speed rail (HSR) is not only more energy efficient than most other modes – even with all life-cycle costs considered – but it can also be a viable alternative mode to other more congested and more polluting modes such as automobiles and air. In fact HSR can be the lowest energy consumer and greenhouse gas emitter if it consistently travels at high occupancy and uses low-emission electricity sources such as wind (ITS Berkeley, 2010).

The impacts of global warming are not felt evenly or everywhere, but they can be serious. Flooding in costal and low-lying areas has especially damaging impacts on roads, rail lines and runways. Modified designs and construction of roads and other paved surfaces, railroads, tunnels and drainage channels to make them more resistant to the damaging effects of flooding are especially important, given the long life and capital-intensive nature of such facilities (Black, 1990, 2007). In addition, metals used in rail lines and other guideways, and the construction methods in such facilities, should be modified to make the facilities more resistant to the negative impacts of higher temperatures and their wider variation. Buckling in railroads, for example, can cause serious damage and thus disrupt normal service by lowering safe operating speeds, and in some cases increasing derailment risks. For example, Samoilovich (2012) demonstrated the possibility of a radical reduction in rail buckling with a choice of rational cooling regimes. Equally important is the early and reliable detection of buckling in rails. Because of the significant safety and operational impacts of buckling, different techniques have been researched, tested and in some cases implemented (Phillips et al., 2010; Signore et al., 1997). The relevance of these and similar sensing and detection methods will become only more critical as temperature increase due to global warming becomes more of a threat. New types of specially treated steel can significantly reduce the occurrence of such problems to start with.

Air-pollution-absorbing concrete and bricks may sound fictitious as means to help clean the air, but it may not be so for long. Such technologies and materials are available, and being used in pilot studies (Shibata et al., 1998). At least one pilot study was reported in the Netherlands, where pollution-absorbing bricks were used to pave the road. While widespread application of such material is still a thing of the future, the possibility and potential are both intriguing and encouraging. Transportation noise pollution, particularly from road traffic, is another serious problem that can be reduced using special materials and construction methods for road surfaces, where road–tire interaction is a major source of the road noise (Hamet and Klein, 1994; Meiarashi et al., 1996). Simple measures such as tining of road surfaces (longitudinal tining in particular) were shown to reduce noise levels (Wayson, 1996). Other simple measures that do not require new practices or materials, and which do not compromise friction (and hence safety) are also available and can be used to reduce roadway noise levels (Rasmussen et al., 2008).

Constructing transportation facilities can be done sustainably and with minimum impact on natural habitat. A comprehensive account of all such methods is not the objective here, but a few examples and considerations demonstrate how alternative construction methods and practices can go a long way towards fostering transportation sustainability. Bridges, tunnels and culverts are necessary components of large transportation construction projects. These components can be constructed in a way not to disrupt wildlife migration routes (Ward, 1982; Klein, 1971; Verkaar and Bekker, 1991), fish passage and nesting grounds (I-Last, 2010). In addition, necessary mitigation of damage by such projects, when this is not preventable, should be undertaken so as to guard against fragmentation of natural habitats. In the rating system of I-LAST, measures of avoidance, minimization and mitigation of fragmentation of natural habitats were explicitly noted as indicators of project sustainability.

Last but not least, necessary policies and supporting tools need to be in place to facilitate and promote sustainability in transportation. For example, to help cities attain more sustainable urban transport systems, the European Commission launched the City–Vitality–Sustainability (Civitas) initiative. Civitas helps cities implement, demonstrate and evaluate an integrated mix of technology and policy-based measures (Sustainability for Road Infrastructure, 2010). While the initiative is not construction oriented per se – it is more policies and technologies oriented, but all, at some level, and to a varying extent, have direct links with and impacts on construction activities – its structure, implementation options and flexibilities, and successes are worthy of note. The rating system adopted by the Illinois Department of Transportation (I-LAST, 2010) is a good example of a transportation agency initiative to promote and systemize the adoption of sustainability in transportation projects.

EXAMPLE CASE STUDY

The example case study used in this chapter is Beheiry and Abu-Lebdeh's (2009) development of the Sustainable Construction Technology Index (SCTI) and its use to measure sustainable technology use in commercial and residential projects executed in the United Arab Emirates over the time period 2004–09. The SCTI was created by accumulating sustainable technology indicators from the literature, and validated by construction industry experts.

The SCTI was built on a Likert scale to assess the degree of use of sustainability tools at every phase of the project. The scale ranged from (5) technology used to the maximum possible benefit of the tool, to (4) technology used extensively, (3) technology used moderately, (2) technology used on a limited basis, (1) technology considered but not used, and (0) technology not considered. The tool encompassed five major technologies in each phase of the project. The total possible score for the index was 100 percent (25 percent for total use of each of the four phases).

The SCTI tackled the four stages of building construction projects (planning, design, construction and occupancy). The planning phase included tools such as the Green Project Management Process (GPMP), green space preservation, public transport, sustainable urban infrastructure and walkable communities. The design phase considered the use of environmental modeling to minimize damage to the natural environment, design for reuse and recycling, design impact measures, low-impact materials and biomimcry, and low and renewable energy systems. The construction phase involves the use of reusable modular formwork, clean

Table 10.1 Technology use

	Residential projects	Commercial low rise	Commercial high rise	All data set
N	69	27	14	110
Total possible score	100%	100%	100%	100%
Average total score	20	28	41	29.6%
Total possible score per phase	25%	25%	25%	25%
Planning average score	2	5	10	5.6%
Design average score	9	10	14	11%
Construction average score	3	4	7	4.7%
Occupancy average score	4	5	10	6.3

Source: Beheiry and Abu-Lebdeh (2009).

sites methodology, low-energy building techniques, solar-powered devices and low interference with the surrounding habitat. The occupancy phase illustrates the implementation of the previous phases in natural light and ventilation use, utility use reduction and the building's integration with the natural environment.

The validation data sample was 110 projects (Table 10.1). These comprised 69 residential buildings and 41 commercial buildings. The data were collected by contacting project managers working for owners only. The data examined the total SC technology use per phase of construction, and the particular differences between the phases. Low-hanging fruit for quick improvements in future technology use was also identified for the different categories.

This pilot study of the SCTI for residential and commercial building projects was followed up by further data examination study (Beheiry, 2011), where 208 projects were collected (130 residential condos, villas and apartment buildings; 51 commercial low-rise buildings; and 27 commercial high-rise buildings). The studies showed that commercial high-rise buildings took more note of sustainable construction principles and attempted to use available technology more often than other building projects.

REFERENCES

Acquaye, A., A. Duffy and B. Basu (2009), 'Embodied energy analysis: a sustainable construction design assessment tool', Conference Paper, Dublin Institute of Technology.

Augenbroe, G. and A. Pearce (2012), 'Sustainable development in the United

States – a perspective to the year 2010', CIB-W82 Report, Georgia Institute of Technology.

Beheiry, S.M. (2011), 'Benchmarking sustainable construction technology', *Journal of Advanced Materials Research*, **347–353**: 2913–20.

Beheiry, S. and G. Abu-Lebdeh (2009), 'A framework for benchmarking sustainable technology use', *Proceedings of the 2nd International Conference MESD'09 on Strategies for Sustainable Technologies and Innovations*, France.

Black, W.R. (1990), 'Global warming: impacts on the transportation infrastructure', *TR News*, **150**: 2–8.

Black, W.R. (2007), 'From global warming to sustainable transport', *International Journal of Sustainable Transportation*, **1**(2): 73–89.

Boyko, C.T., M.R. Gaterell, A.R.G. Barber, J. Brown, J.R. Bryson, D. Butler, S. Caputo, M. Caserio, R. Coles, R. Cooper, G. Davies, R. Farmani, J. Hale, A.C. Hales, C.N. Hewitt, D.V.L. Hunt, L. Jankovic, I. Jefferson, J.M. Leach, D.R. Lombardi, A.R. MacKenzie, F.A. Memon, T.A.M. Pugh, J.P. Sadler, C.Weingaertner, J.D. Whyatt and C.D.F. Rogers (2012), 'Benchmarking sustainability in cities: the role of indicators and future scenarios', *Global Environmental Change*, **22**(1): 245–54.

Dunay, R., Wheeler J. and R. Schubert (2006), 'No compromise: the integration of technology and aesthetics', *Journal of Architectural Education*, **60**(2): 8–17.

Evans, D.L. (1997), 'High-luminance LEDs replace incandescent lamps in new applications', *Proceedings SPIE 3002*, p.142.

Garnett, N. and S. Pickrell (2000), 'Benchmarking for construction: theory and practice', *Construction Management and Economics*, **18**: 55–63.

Hazelton, P. and M.A. Clements (2009), 'Construction of an environmentally sustainable development on a modified coastal sand mined and landfill site – Part 1. Planning and implementation sustainability', *MDPI – Open Access Publishing*, **1**(2): 319–34.

Hill, R.C. and P.A. Bowen (1997), 'Sustainable construction: principles and a framework for attainment', *Construction Management and Economics*, **15**: 223–39.

Hoa, T., G. Qi, Y. Yan and H. Yang (2011), 'Reuse of reclaimed materials in construction- from the process analysis of BedZED', *Applied Mechanics and Materials*, **99–100**: 433–9.

Horsley, A., C. France and B. Quatermass (2003), 'Delivering energy efficient buildings: a design procedure to demonstrate environmental and economic benefits', *Construction Management and Economics*, **21**: 345–56.

Hwang, B.G., S.R. Thomas, D. Degezelle and C.H. Caldas (2008), 'Development of a benchmarking framework for pharmaceutical capital projects', *Construction Management and Economics*, **26**: 177–95.

I-LAST (2010), 'Illinois – livable and sustainable transportation rating system and guide', Illinois Department of Transportation.

Khasreen, M.M., P. Banfill and G. Menzies (2009), 'Life-cycle assessment and the environmental impact of buildings: a review', *Sustainability. MDPI – Open Access Publishing*, **1**(3): 674–701.

Kibert, C.J. (2003), 'FORWARD: sustainable construction at the start of the 21st century', *Journal of the Future of Sustainable Construction*, Special Issue: 1.

Kibert, C.J. (2008), 'Green building education and research at the University of Florida', *Associated Schools of Construction Annual Conference – International Proceedings*, available at http://ascpro.ascweb.org/chair/paper/CEGT183002009.pdf.

Klein, R.D. (1971), 'Reaction of reindeer to obstruction and disturbances', *Science*, **173**(3995): 393–8.

Mehta, R. and L. Bridwell (2005), 'Innovative construction technology for affordable mass housing in Tanzania, East Africa', *Construction Management and Economics*, **23**: 69–79.

Meiarashi, M., M. Ishida, T. Fujiwara, M. Hasebe and T. Nakatsuji (1996), 'Noise reduction characteristics of porous elastic road surfaces', *Applied Acoustics*, **47**: 239–50.

Ochoa, L., R. Ries, H.S. Matthews and C. Hendrickson (2005), 'Life cycle assessment of residential buildings', *Proceedings of the Construction Research Congress: Broadening Perspectives*, San Diego, CA, pp. 87–91.

Phillips, R., I. Bartoli, S. Coccia, F.L. d Scalea, S. Salamone, C. Nucera, M. Fatech and G. Carr (2010), 'Nonlinear guided waves in continuously welded rails for buckling prediction', *AIP Conference Proceedings*.

Prowell, B.D. (2007), 'Warm mix asphalt. The international scanning program', Summary Report, American Trade Initiatives.

Rasmussen, R.O., S.I. Garber, G.J. Fick, T.R. Ferragut and P.D. Wiegand (2008), 'How to reduce tire-pavement noise: interim better practices for constructing and texturing concrete pavement surfaces', Concrete Pavement Technology Center, Iowa State University.

Richardson, B.J.E. and D.O. Jordan (1994), 'Use of recycled concrete as a road pavement material within Australia', *Proceedings of the 17th ARRB Conference*, Queensland, **17**(3): 213–28.

Samoilovich, Y.A. (2012), 'Possibility of producing railway rails with increased strength and minimum buckling', *Metallurgist*, **55**(11–12): 903–11.

Qian, Q. (2008), 'Strategies of implementing a green building assessment system in Mainland China', *Journal of Sustainable Development*, **14**(1): 2.

Signore, J.M., M.G. Abdel-Maksoud and B.J. Dempsey (1997), 'Fiber-optic sensing technology for rail-buckling detection', *Transportation Research Record*, **1584**: 41–5.

Steinert, J. (2008), 'Developing world sustainable building practices: a look at buildings in impoverished locales', Master's report in Construction Engineering and Management, University of Colorado at Boulder.

Toutanji, H.A. (1996), 'The use of rubber tire in concrete to replace mineral aggregates', *Cement and Concrete Composites*, **18**(2): 135–9.

Transportation Research Board (TRB) (2004), 'Integrating sustainability into the transportation planning process', *Conference Proceedings 37*.

UN (1987), *Development and International Economic Co-operation: (Report of the World Commission on Environment and Development): 'Our Common Future'*, Official Records of the General Assembly, 42nd Session, Supplement No. 25.

UNEP (2002), 'Agenda 21 sustainable construction in developing countries', United Nations Environment Programme Publications and Reports.

UNEP (2002), 'Industry as a partner for sustainable development', United Nations Environment Programme Publications and Reports.

UNEP (2003), Halls, S., The UNEP International Environmental Technology Center, United Nations Environmental Program Publications and Reports.

Verkaar, H.J. and G.J. Bekker (1991), 'The significance of migration to the ecological quality of civil engineering works and their surroundings', *Nature Engineering and Civil Engineering Works*, issue 44-61.

Walubita, L.F., G. Das, E.M. Espinoza, J.H. Oh, T. Scullion, S. Nazarian,

I.Abdallah and J.L. Garibay (2011), 'Texas flexible pavements and overlays: data analysis plans and reporting format', Texas A&M University, Report 0-6658-P3.

Wayson, R.L. (1996), 'Relationship between pavement surface texture and highway traffic noise', National Cooperative Highway Research Program, Synthesis 268.

Ward, A.L. (1982), 'Mule deer behavior in relation to fencing and underpasses on Interstate 80 in Wyoming', *Transportation Research Record*, **859**, 8–13.

Yigitcanlar, T. and F. Dur (2010), 'Developing a sustainability assessment model: the sustainable infrastructure, land-use, environment and transport model', *Sustainability. MDPI – Open Access Publishing*, **2**(1): 321–40.

Web References

Bynum, P., R.R.A. Issa and S. Olbina (2012), 'Building information modeling in support of sustainable design and construction', *Journal of Construction Engineering and Management*, available at http://ascelibrary.org/coo/resource/3/jcemxx/423 (accessed 30 March 2012).

COLAS Group (n.d.), available at http://www.colas.com/en/responsible-development/other-challenges/energy-and-greenhouse-gases/energy-content-of-products-offered-to-customers-940147.html (accessed 01 June 2012).

Dewlaney, S.A., M.R. Hallowell and R.B. Fortunato (2011), 'Safety risk quantification for high performance sustainable building construction', *Journal of Construction Engineering and Management*, available at http://ascelibrary.org/coo/resource/3/jcemxx/373 (accessed 26 October 2011).

Hamet, J. and P. Klein (1994), 'Road texture and tire noise' available at http://www.inrets.fr/ur/lte/publications/publications-pdf/web-hamet/in00_674.pdf (accessed 18 February 2012).

ISTEA (1991), 'Intermodal Surface Transportation Efficiency Act of 1991', *USA Public Law 102-240*, available at http://ntl.bts.gov/DOCS/istea.html (accessed 25 May 2012).

ITS Berkeley (2010). Tracking high-speed rail's energy use and emissions', Berkeley Transportation Letter, Spring, available at http://its.berkeley.edu/btl/2010/spring/HRS-life-cycle (accessed 1 June 2012).

Korkmaz, S. and A. Singh (2011), 'Impact of team characteristics in learning sustainable built environment practices', *Journal of Professional Issues in Engineering Education and Practice*, available at http://ascelibrary.org/epo/resource/3/jpepxx/57.10 (accessed 7 December 2011).

Pennisi, E. (1992), 'Rubber to the road – recycled tires become asphalt', *Science News*, available at http://findarticles.com/p/articles/mi_m1200/is_n10_v141/ai_12033403 (accessed 1 June 2012).

Shibata, H., M. Satokawa, Y. Im and K. Lim (1998), 'Method for manufacturing carbon dioxide absorber using concrete sludge', available at http://www.freepatentsonline.com/EP0610781.html (accessed 17 February 2012).

Sustainability for Road Infrastructure (2010), 'Strength in numbers – a collaborative approach to developing sustainable transportation', available at http://www.ropl-digital.com/mags/sustRI1012/sustRI1012.pdf (accessed 1 June 2012).

11. The eco-logistics improvement in France: towards a global consideration of inland waterway transport within the supply chain strategy

Thierry Houé and Renato Guimaraes

STRATEGIC, COMPLEX, INTEGRATED AND SUSTAINABLE SUPPLY CHAINS

Nowadays, sustainability strategies of companies are particularly important for all economic activities. Understanding the origin of economic and environmental problems of each sector and organizational function can explain how to develop a real sustainable value strategy for many organizations (Orsato, 2006, 2010). It is probably one of the best ways to serve customers and to ensure a long-term relationship with them. It is also the solution to improve the profitability and the sustainability of the firm with a powerful competitive advantage (Sullivan, 2006). Freight transport operations are affected by a wide range of strategic decisions made at both a logistical and a corporate level. Of course, these decisions influence various characteristics of transport operations. In consequence, the definition and the implementation of sustainable strategy have a significant effect on the transport policy of the company and in particular on the choice of transport mode. Based on the fact that the main objective of this chapter is to highlight issues related to the implementation of inland waterway transport (IWT) solutions by companies, we thought it necessary to clarify often misinterpreted concepts. We explain and redefine the concepts of sustainable development and sustainable logistics. Defining these concepts enables us to define the scope of this study. In a next step, we explain that the development of a multitude of supply chain networks leads to more complex organizational situations that firms must adapt to in order to guarantee the performance of their flows.

More Integrated and Sustainable Logistics

First of all, we must clarify what the concept of sustainable development really is. This concept was first introduced in the Brundtland Report of the World Commission on Environment and Development (Brundtland, 1987). Sustainable development is then defined as development that meets our present needs without compromising the ability of future generations to meet their needs (Schmidheiny, 1992). It is based on the idea that it is possible to create values for the firm, people and the planet. Basically, this is a holistic attitude that takes into account environmental, societal and economic issues. The company is perceived as an entity operating in a society and more specifically in an environment in which it develops a close relationship with its environment (Dewberry, 1995). This is a fundamental approach that all kinds of organizations must now integrate into their modus operandi. Faced with organizational, functional, strategic and environmental constraints, companies are now under pressure to adopt more responsible practices for managing people and resources.

Originally compartmentalized and fragmented within organizations, logistics was gradually transformed into a strategic and integrated mesh (Bowersox and Closs, 1996; Clinton and Closs, 1997) serving business agility (Christopher, 2000). In the last 20 years, this function has become a tool serving the customer orientation strategy for many companies. The main reasons for this metamorphosis are strategies of concentration, the development of lean manufacturing, the growing pressure of competition and the cost reduction effort (Christopher, 2011). Besides, thanks to the arrival of business process re-engineering (BPR) in the 1990s (Hammer and Champy, 1993), the relationship between logistics and the other parts of the firm was redefined. BPR detects core processes, crossing traditional functions boundaries. They are fundamentally customer-oriented. Effective management of these processes requires new working methods and the improvement of relationships between functions (McKinnon, 2001). Step by step, the word 'logistics' was gradually replaced by the concept of the supply chain which enriches it, emphasizing a vision of interorganizational coordination of economic activities (Mentzer et al., 2001; Mentzer et al., 2008). This comprehensive approach is similar to the concept of sustainable development which, as mentioned above, can be seen in terms of interactions between societal, economic and environmental areas. Figure 11.1, inspired by the work of the NSTC (National Science and Technology Council, 2001) in a report on the integration of transport systems in 2050 and a study on key technologies for 2010 by the French Ministry of Economy (MEIE, 2006) shows the three aspects of sustainable logistics. It also specifies the clear interactions between each of these dimensions.

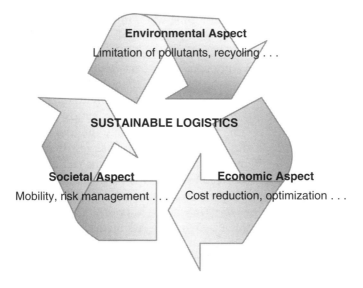

Source: Adaptation from National Science and Technology Council (2001)

Figure 11.1 The three dimensions of a sustainable logistics

To illustrate how the three dimensions are implemented in a decision-making process, we take as an example the choice of a more eco-friendly mode of transporting freight. This can have a direct influence on the environmental sphere. In addition, this decision may also impact on costs, thus optimizing the supply chain (Beamon, 1999). It may also affect the way society works in general or, more precisely, the safety of transportation or even consumer behaviour.

The Multiplicity of Logistics Networks: The Real Source of Complexity

Regarding global supply chains of companies, they are formed by very diversified and interconnected networks of flows. Moreover, the 'chain' metaphor seems less and less adapted to this new situation (Harland et al., 2001). These multiple logistics networks inevitably have an influence on the steering and coordination of supply chains. According to Fisher (1997), the supply chain strategy should be based on the product type (often due to the diversification of activities and products). The nature of goods can lead companies to prefer one transportation mode over another. These organizational changes have consequences both for the evolution of the structure and for the transportation strategies applied by the companies, especially at a time when they are increasingly taking into account the various aspects of sustainable development (Maxwell and Van der Horst, 2003). The integration of ever more players into the heart of the supply chain (suppliers, third-party logistics, shippers, retailers, etc.) considerably increases the links that generate more complex relationships between the partners. Therefore, many trading systems can appear and may take different forms ranging from simple outsourcing to genuine collaborative relationships. A challenge to the traditional market relationships between firms accompanies such changes. It implies a sustained coordination of physical flows by information flows, requiring greater interaction between the company and its partners (Gunasekaran and Ngai, 2004). From this point of view, it is preferable to consider the supply chain as a set of networks in which a huge diversity of links between the various players is materialized and players themselves become part of wider networks. The operating level of the logistics networks is characterized by a mesh of transport and information systems infrastructures, and patterns of flow organization. A more strategic view of these networks complements the operational representation. This has gradually become clear with the continued strengthening of cooperation, alliances and partnerships between stakeholders (Christopher, 1999; Lambert et al., 2006). The great diversity of relationships and movements of goods and information flow also give a strong spatial and temporal characteristic to these forms of organization. The supply chain is no longer symbolized solely by a set of basic transactions but evolves into a structure in which relations between partners, time and space are intermingled. It then goes without saying that transport strategies pursued by firms will inevitably be influenced by the complexity of this network. This is embodied by the use of alternative modes of transport rather than 'just the road' (Savy, 2007).

All major companies agree that the problem of transporting goods is probably the most important logistics challenge for the next ten years. Transportation effectively has an impact on the three pillars of sustainable development: the societal aspect, with its effect on the mobility of people and goods, or diseases caused by pollutant emissions from trucks; the environment aspect, as the sector producing the highest greenhouse gas (GHG) emissions with 28 per cent of gross emissions in France (ADEME, 2008); and the economic aspect with its serious impact on growth, trade, overall costs and the ongoing search for optimization (Savy, 2007). The transportation sector alone represents nearly 70 per cent of final energy consumption of oil products in France. Companies are studying different ways to tackle these problems. One possibility is to work directly on the nature of the vehicles. Some logistics service providers, such as UPS and FM Logistic, are trying to renew and preserve their regular fleet of trucks or planes (Schmidt, 2005). The large industrial shippers such as Arcelor-Mittal, Sanofi-Aventis and PSA, and groups of large retailers such as Carrefour, are proposing to review their supply chains completely by rationalizing the number of sites and warehouses and improving vehicle loading rates. But the players are also looking for modal or multimodal cleaner alternatives such as waterways, rail or combined rail–road transport which offer better fuel efficiency than road-only (Becker, 2003). In July 2008, Auchan, Casino, Carrefour, Conforama, Ikea and Leroy Merlin subscribed to the 'great river transport and distribution' protocol applicable to the Rhône-Saône: the aim is to commit to focus on transportation by barge for the delivery of non-food products. Estimates show that the agreement should prevent the emission of nearly 1500 tonnes of CO_2 per year (Préfecture de Région Rhône-Alpes, 2008).

THE REVIVAL OF FRENCH INLAND WATERWAY TRANSPORT

The contemporary history of the French river system begins in the nineteenth century. To meet demand and to resist the expansion of the railways, two men were working on developing the river system. Becquey Louis, director of the famous 'Ponts et Chaussées' School at the beginning of the nineteenth century, initiated the action that would create a network of waterways. He established a system of innovative financing for the construction of new channels and the standardization of existing channels to a minimum gauge. Charles-Louis de Freycinet, Minister of Public Works from 1877 to 1879, had meanwhile established a plan to encourage the development of French heavy industry by improving transport conditions. This plan was responsible for modernizing and expanding the existing network. The law of 5 August 1879 gave France a homogeneous network of waterways. More than 700 million francs were invested at the time. Soon, 2453 kilometres of roads were upgraded and 468 kilometres of new canals were built. The Act brought with it other changes, such as unifying the network by creating a unique gauge for the main waterways. It was also responsible for standardizing the boats to a length of 38 metres. Twenty years after the entry into force of this Act, the main network of waterways exceeded 4739 kilometres. Unfortunately, it would suffer greatly due to the First World War. All proposed projects would be abandoned (such as the construction of a link from the Mediterranean to the North Sea and the Rhine). The rise of the automobile combined with political support for rail transportation led to government disinterest in the waterways and of course a reduction in public aid. The Second World

War would destroy almost 20 per cent of the fleet of barges operating in 1938. The infrastructure became neglected. The network of nearly 8500 km is mainly composed of small canals (Freycinet gauge), with a gauge not adapted to business needs. The flows of goods increased. But apart from the two world wars, the decline of IWT also had other causes: changes in economic structure with the decline of heavy industry; the obsolescence of the network; lack of maintenance resulting in operational difficulties; rigid practices; obsolescence of the legislative and regulatory framework; and strong competition from other modes of transport, including road transport in the 1970s.

However, since the mid-1990s, the waterways have been experiencing a real resurgence of interest and upgrading in France. It should be noted that this period of growth was preceded by the establishment, in 1991, of the French agency for the inland waterways. Named the Voies Navigables de France (VNF), this public institution, attached to the Ministry of Ecology and Improvement of Sustainable Development, has several missions. The most important is the management, modernization and development of 6700 kilometres of the inland waterways. VNF federates the initiatives on transport, but also tourism. VNF operates a public area of nearly 80 000 hectares across 57 administrative departments. It relies on the expertise of its staff and its 4500 agents. This is significant and shows that in France, transport policy, particularly in terms of rivers, was and still is strongly influenced by the state. It is virtually impossible to imagine a mode switch without strong encouragement from the public authorities. This is important and shows the institutional world as a major player in the development of companies' transport strategies.

Inland Waterway Transport which has been Increasing in France for Ten Years

The use of the inland waterway on French territory has increased in ten years by about 40 per cent in tonnes/kilometre. From 5700 tonnes/ kilometre in 1996, it rose in 2006 to nearly 8000 tonnes/ kilometre, a growth of 3.3 per cent per year (Figure 11.2). Despite a decline in 2008 (due to the economic crisis), these results are particularly remarkable on wide-gauge waterways, where mass flows of goods are possible. The Rhone-Saône canal has almost doubled its traffic since 1998. The Seine River and the Nord-Pas de Calais Region showed an increase of almost 50 per cent. Foreign achievements, and primarily the development of the wide-gauge canal linking the Rhine to the Danube, that opened in 1992, have launched the debate on development of IWT. Wide-gauge waterways such as the Seine, Rhone, Moselle and Rhine rivers still have a reserve capacity ena-

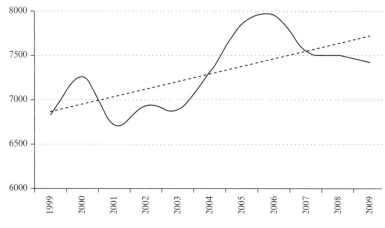

Source: VNF (2010)

Figure 11.2 *The evolution of IWT in France from 1999 to 2009 (million tonnes/km)*

bling them to absorb volumes of traffic well above current levels, without recourse to additional infrastructure work (VNF, 2008).

Cyclical and Structural Factors behind this Increase

This development of inland waterways is caused by structural and cyclical changes. From an environmental perspective, the world is in turmoil. Awareness of climate change, increased pollution, increased traffic on the road, noise and odour, require manufacturers to seek new alternatives to 'just the road'. The economic aspect is obviously important. Despite a recent but temporary decline in the price per barrel of crude at around $70 in September 2009, the regular and inevitable rise in oil prices will eventually force companies to rethink their transport strategy and find ways to reduce their overall logistics costs. In terms of timeliness and quantity, firms are now forced to use more efficient modal solutions. Changes to the legislative and regulatory framework are also the source of these changes. The Act of 12 July 1994 and its decrees may evolve in a liberal direction. It abolished the trading rules of 1941 without advocating ultra-liberalism. It opens the market to competition and allows the relaxation of volume contracts by opening them to all. It translates into real sector accountability. The Act of 16 January 2001 adapts French law to European Community law. Regulated tariffs have been removed. All these points are also sources of resurgence in demand for river transport.

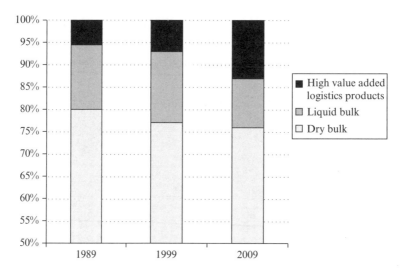

Source: VNF (2010).

Figure 11.3   Types of goods carried by IWT in France (based on tonnes/ km performed)

While this mode has been known and appreciated for many years for handling heavy goods, it tends to be increasingly used for transporting high-value-added logistics goods (Figure 11.3). In addition to the sharp increase in traffic on inland waterways, a structural change has been recorded in the nature of goods transported. In ten years the share of high-value-added goods has risen steadily. The weight of this market (mainly containers) in the river traffic has tripled in ten years and reached 7.8 per cent in 2008. At the same time, the dry bulk market well known to the watercourses fell from 81 per cent t-km in 1989 to 75 per cent of traffic in 2008. This development is not without consequences. It preserves the river mode from hazards encountered on the markets closely linked to conditions such as the grain market or that of energy products.

The combined river–road mode shows handsome growth of around 30 per cent and now accounts for one-third of combined transport in France. The growth of shipping is due mainly to higher levels of freight in agro-food, construction materials, petroleum products and coal. Many initiatives are also behind this revival. The port of the town of Lille has created and operates a route for used glass by waterway between the Triselec platform of Halluin and a plant located at Wingles. This action represents more than 3500 trucks removed from one of the most congested areas in the north of France. A wide-gauge canal project that could disrupt the logistical schemes of Northern Europe has entered a decisive stage. This is the Seine Nord Europe Project, a section of waterways 105 km long and 60 metres wide, being dug next to the old Canal du Nord through Picardie, between the Oise and Compiegne and the Dunkerque Escaut canal near the town of Cambrai. This infrastructure will bring the ports of Le Havre and Rouen out of isolation, as well as the Ile-de-France region and the Seine area that remains impassable for large boats. Northern Europe can be reached with convoys that can carry at once the equivalent of four trains and 220 trucks, while the unit costs of transport should be halved for shippers. Another often overlooked advantage of waterways is that they reach the heart of major cities without influencing the transport of people, unlike road and rail as they face the difficulty of finding new paths.

THE EFFECTS OF AN INTEGRATION OF INLAND WATERWAY WITHIN SUPPLY CHAINS

In this section, we look at the development of logistics and its effects on transport policies and infrastructures. Then we consider the geographical aspect of collaborative logistics networks. Finally, we show the consequences of the integration of inland waterways into the logistics networks of companies in France. We propose to address both factors facilitating integration and difficulties related to the use of the waterways.

The Role of Transport and Territorial Policies for the Development of Logistics

Business growth cannot be possible without a joint improvement of infrastructure. This is what the economists suggest. Transport infrastructure is a means of development, a way to reduce transport costs, to improve accessibility and to motivate the location of firms (Offner, 1993). In their spatial dimension, the economic and technological developments have an importance in the field of goods transportation. Elements such as inventory reduction along the production and distribution chain, flexible production techniques or the pressure of demand, reinforce the interest in transport and logistics functions (Savy, 1995). The use of complex logistics solutions can be explained by the movement of skill diversification facing logistics companies, a movement which tends to encourage shippers to use the services of a third party for logistics (Elram and Cooper, 1990; Mortensen and Lemoine, 2008). For several years, with the White Paper on European Transport Policy for 2010 (European Commission, 2001), this debate has intensified, and has recently materialized in France with the production of several reports and numerous studies, the proposals in which have fuelled thought. Recent decisions will shape the landscape and transport infrastructures for 2025 (CIADT, 2003), will define the commitments of the French state and will naturally have very strong implications for the development of logistics activities in all regions of the country. In France, transport infrastructure is usually managed by the government. In addition to contributing to economic development, it is a major instrument of regional planning. The main orientations of current thinking are part of a policy aimed at:

● improving the attractiveness of France by strengthening the international openness of the country (ports, hubs, platforms, and so on);
● developing logistics in France (new corridors, quality of logistics platforms);
● improving the accessibility of large French cities;
● considering the improvement of transport for sustainable development and as far as possible increasing the modal shift from road to other types of transport.

As part of a European dimension, these guidelines are divided by major project and region. However, new infrastructure alone cannot ensure the economic development of a region (Meunier, 1999). It is therefore necessary to adapt behaviours and the strategies of economic players to understand the key factors to sustain or encourage the development of logistics activities.

One of the objectives of logistics is cost reduction combined with improved service levels. Lean management is now practiced by many companies. An ever more demanding application requires firms to modify their production, their supply and distribution systems, and therefore the entire structure of their logistics flows. This is imperative for the competitiveness of industrial shippers, distributors and logistics providers. This finding is far from being without effect on the choice of transport mode (McKinnon, 2001). It is clear that the road transport mode appears, for the moment and for many reasons, to be the most efficient to meet these constraints. Integrated into the discussion on development of transport infrastructures, these elements show the antagonism between the specific objectives of logisticians seeking to cumulate their flows to benefit from economies of scale, and those trying to maintain a certain balance in spatial development (Dornier and Fender, 2007).

The Collaborative Aspect and Geographical Dimension of Logistics Networks

The establishment of logistics networks is reliant on changing relationships between companies. This started 30 years ago. The restructuring of firms has accelerated and involves the development of more collaborative structures. To overcome the drawbacks of centralized organizations, these forms are now oriented towards cooperation (Batsch, 1993). Increased competition is characterized by increased accessibility to markets and interconnections between people (Miles et al., 1992). But the importance of time in global competition (Stalk and Hout, 1992) explains the causes of these changes in organizational forms. The network structure has contributed to a faster reorganization of relationships in the supply chain. Through their transactional characteristics (with outsourcing or the integration of logistics information systems) and relational characteristics (with relations between players within networks), it is useful to talk about a plurality of networks to describe the reality of the supply chain. The logistics networks are at the heart of a spatial phenomenon of geographic integration systems where transport and communication play a leading role (Melo et al., 2009). This geographic integration materializes in the appearance of national and global logistics activities management, combining principles of organization, information systems and infrastructures. Seeking this integration means that companies must go beyond geographic boundaries to create new zones of operation and to manage the flow issues. This integration challenge, based on the efficiency of supply chains, economies of scale and standardization, requires suppliers to consider the location of production units and distribution centres in limited places at an international level (Paché, 2006). This integration combines geographic aspects with the deployment of new organizational forms.

The Integration of Inland Waterway Transport in Logistics Networks

The advantages of inland waterway transport for a logistics network are various. They facilitate traffic flow, provided that this mode is integrated into a comprehensive transport management system.

Towards lower transport costs in the logistics network

IWT provides economic and ecological benefits. This is also a safe transportation mode that often limits hidden costs. It is of economic interest in terms of energy consumption in a logistics network. With only 1 gallon of fuel, 1 tonne of freight travels 6.6 km by air, 100 km by road, 330 km by railroad but 500 km by waterway. All internal costs combined, the cost per tonne/km for a barge convoy is €0.012 to €0.015 compared with €0.015 to €0.042 for a truck and from €0.0255 to €0.03 for a train. Currently, a 20 foot container shipped from Rotterdam to Strasbourg costs €750 by railway, €720 by road and €360 by waterway. Boats are habitually slow. But the economic advantage mentioned above is not cancelled out as long as profitability is not expressed in terms of speed of movement but in travel time availability. Appropriate infrastructures allow boats to navigate every day. Rotterdam–Strasbourg can be done in only 48 hours, Paris–London in two days, Paris–Bilbao in four days and Paris–Casablanca in seven days.

An ecological benefit for the network and its image
Over the past decade, ecological regulations have intensified. The Kyoto Protocol, European Union (EU) directives, decrees regarding recycling wastes or metropolitan initiatives to protect the environment – all these new demands require companies to react quickly and determine the structuring of their transport networks and their modal choices. They also require them to manage their communication (branding) to their customers. The mission of logistics is now to reconcile the company with its natural environment. It is time to discuss efforts to implement and to combine new resources within the supply chain. The challenge of eco-logistics seems both a necessity and an opportunity, especially if firms consider it as a vector of performance (Ummenhofer, 1998). IWT fits in perfectly with this movement. It is friendly to energy resources, the environment and people. Germany is one European country in which the green movement is particularly extensive. This country has already spent nearly 45 million euros on investment in this transport mode. The waterways save fuel per tonne-km transported, thereby preserving energy resources and reducing gas emissions at a ratio of 1:5 by comparison with road transport. The noise of trucks and planes causes many disturbances: psychological and social problems, as well as low safety levels (for road only). These disorders depend on whether the background noise is during day or night. Noise tops the list of perceived nuisances. The waterways produce very little noise (a peak is located below 70 decibels) while road transport generates a sudden and long sound shock, which is particularly aggressive at night. Thanks to the waterways, 4000 tonnes of goods cross the city of Lyon in just 15 minutes. Consequently, even in the context of a sustained increase in traffic, noise production by a boat is and will remain insignificant. This mode has the energy advantage characteristic of heavy transport modes. Compared to the costs of pollution of road or rail damage, IWT is more ecological. Indeed, this is the mode that offers the highest tonnage per km. Thus, it is more economical in space. Finally, in terms of integrating ecological and aesthetic aspects, it

is possible to build a network while respecting the environment. Examples can be cited, such as the Allant River and the Saône River close to the city of Macon, featuring an irregular shoreline and wooded islands inhabited by birds, and so on.

Inland waterway transport: a mode to reduce logistics risks and external costs

Safety is the major competitive advantage of inland waterway transport. Compared to other types of transport, IWT is one of the safest ways of transporting goods and, of course, people. Each year, about 4000 people are killed on the roads in France. This figure has financial consequences for firms. The increased numbers of trucks on the roads boosts the potential accident risk. In terms of accidents, the level of risk for the waterways is very low. The main risks relate to any problems that may occur during transportation of hazardous materials which can then affect people and the environment. But with this kind of transport, this probability is almost zero. As a comparison, the number of road accidents involving hazardous materials is 200 per year on average in France. In addition, road and rail transportation bring together hazardous materials, passengers and residents. Some ecologists suggest taking into account quantitative but also qualitative aspects in order to reflect the true cost of transport. They demand that the various external costs, including the cost of destruction at the end of life cycle of the product, be included in the price charged to the customer. This qualitative view, also called 'green balance', takes into account the cost of accidents, air pollution, noise and climate change. However, it ignores ground pollution, water pollution, and so on, which would increase costs. Current studies on external costs for transport activities have essentially concerned road transport. In fact, road transport has the largest share in total external costs of transport. In order to cover all transport modes and to understand differences and similarities on external cost estimation from one mode to other modes, certain elements have to be considered (Maibach et al., 2008). Table 11.1 provides an overview of the most important specification of different costs according to transport modes in Europe. It shows that IWT allows for minimizing the external costs, at least compared to road transport.

A Transportation Mode Dedicated to a Flexible Production System

The nature of the production system is a determining factor in the choice of transport mode. For many years, IWT has been almost systematically rejected in favour of the road transport. Why? Because many industrial companies had focused on a flexible model of production such as

Table 11.1 Most important specification of different costs according to transport modes

Cost component	Road	Rail	Air	Water
Costs of scarce infra-structure	Individual transport is causing collective congestion, concentrated on bottlenecks and peak times.	Scheduled transport is causing scarcities (slot allocation) and delays (operative deficits).	See Rail.	If there is no slot allocation in ports/ channels, congestion is individual.
Accident costs	Level of externality depends on the treatment of individual self accidents (individual or collective risk) insurance covers compensation of victims (excluding value of life).	Difference between driver (operator) and victims. Insurance is covering parts of compensation of victims (excluding value of life).	See Rail.	No major issue.
Air pollution costs	Roads and living areas are close together.	The use of diesel and electricity should be distinguished.	Air pollutants in higher areas have to be considered.	Air pollutants in harbour areas are complicated to allocate.
Noise	Roads and living areas are close together.	Rail noise is usually considered as less annoying than other modes (rail bonus). But this depends on the time of day and the frequency of trains.	Airport noise is more complex than other modes (depending on movements and noise max. level and time of day).	No major issue.
Climate change	All Green House Gas (GHG) relevant.	All GHG relevant, considering use of diesel and electricity production.	All GHG relevant (Air pollutants in higher areas to be considered).	All GHG relevant.
Nature and landscape	Differentiation between historic network and motorways extension.	Differentiation between historic network and extension of high speed network.	No major issue.	New inland waterways channel relevant.

Source: Maibach et al. (2008).

232

just-in-time. However, waterways transport offers many advantages. It does not suffer from congestion and restrictions on traffic or the disadvantages of railway transport (priority of some trains, strikes, vandalism, etc.). Through efficient planning tools (e.g. with an advance planning system or 'collaborative planning forecasting and replenishment') and an effective coordination between shippers and carriers, goods can arrive at their destination on time. This timeliness is an important asset of inland waterway transport. In addition, the boat can be used as a floating storage area, either before or after a trip. In terms of information management in the logistics network, the recent evolution of navigational equipment has probably contributed to greater integration. Today, there are boats of all types and sizes equipped with the latest communication and navigation technologies (radars, autopilots, global positioning system, electronic navigation maps, etc.). This will meet the expectations of many shippers in terms of traceability and availability of information in the logistics network.

The fourth point is important. IWT can develop only through a true multimodal scheme involving all players in the logistics network (including state and government). Cooperation between players in the network is needed. With a high level of cooperation, integrating various modes of transport will be easier and more efficient.

CONCLUSION: INTEGRATION WITH THE SUPPORT OF NEW STRATEGIES AND PRACTICES

The integration of IWT and its benefits for the circulation of logistics flows can be achieved only if companies (but also the state and government) adopt new practices and offer new solutions. But to accomplish this integration effectively, companies must be aware of and understand the limits of waterways transport before implementing effective organizational patterns in their supply chain. In the final part of this chapter, as a conclusion, we talk about the obstacles to using this type of transport. Then, we present the strategic, commercial and organizational solutions that can be used to develop IWT at the heart of supply chain management.

Barriers Limiting the Integration of Waterway Transport within Supply Chains

Inland waterway transport has some traditional disadvantages. Several major problems can be mentioned, which must be incorporated in advance in the process of construction of the transport scheme. The slow speed of the boats leads to the storage of goods for a long period. Even if the cost of IWT is relatively low, pre- and post-delivery are often expensive. From a technical standpoint, some waterways cannot carry containers: this is a major obstacle for the development of the transport of high-value-added commodities. The waterways network is unevenly located over the European territory: this represents another weakness.

Furthermore, natural events such as major floods can cause significant traffic problems. Frequently in France, between November and March, when rainfall is often heaviest, the level of the Seine rises. This occurred recently in December 2001. Traffic was particularly disrupted for a week. Periods of potential flooding are becoming less predictable, thereby increasing the difficulties of organizing such transport by different players.

Finally, many seaports and river ports have had to invest due to the increase in container transport. These investments increase the overall cost of transport and are often paid for by the administrative regions or the government. These investments also force a rethink of the geography of infrastructures and equipment with transhipment container systems.

The Key to a Successful Waterway Transport Integration Strategy

Our observations lead us to present what appear to be emerging as best practices for the integration of inland waterway transport in the global supply chain. It seems essential for firms to combine the environmental perspective and economic performance. These two dimensions of sustainable development are a part of the strategic decisions of many companies. Plenty of subsidiaries are now exposed to specific directions to protect the environment but also to cut costs. At present, the economic crisis seems to be accelerating this trend.

Being a large company with international stature probably facilitates the integration of IWT within the supply chain. It offers several advantages. The first is having a certain amount of experience in multimodal transport. Major groups are naturally forced to integrate and manage multiple transport types in their transport plans. They use their experience to facilitate and guide their choices but also to ensure better control of their organization and their costs. The second advantage is having links with many partners in logistics (shipping lines, carriers, logistics service providers). Although these relationships are not solely based on cooperative behaviour, the fact of being an important customer of a firm facilitates reorganization of logistics processes. For example, thanks to agreements with shipping companies, Toys R Us, the famous world retailer of toys and childcare articles, successfully negotiated free storage of its products for two to three weeks in the river port of Gennevilliers in France. The activity of Toys R Us is highly seasonal. This buffer stock facilitates homogeneous distribution of the volumes received in one of its central warehouses in France according to need at particular times during the year. This makes a huge contribution to cutting logistics costs (Guillaume, 2009). The use of a barge also gives the benefit of a floating storage area which can save several days of storage (on average between four and five days). It can also limit the risk of accidents and traffic jam problems.

In addition, companies operating in a global competitive market must develop logistics systems combining the environmental aspect, social responsibility and economic performance. They must improve their image for consumers in terms of environmental friendliness. They seek to honour their commitments to environmental responsibility. The whole supply chain is now concerned. The sale of 'green' products requires 'clean' production but also 'green logistics processes' for storage, transport and distribution. The integration of IWT requires the development of new organizations for business.

Towards New Organizational Solutions

The elements of IWT organization and integration change the relationship between players. They form the basis for a renewal of practices between carriers and shippers. Some competing business combinations located near the waterways have begun to appear. These firms tend to work for one reason: they want to organize aggregation of their flows to use IWT at a competitive cost. Furthermore, by moving to a single geographic area, they minimize the constraints of pre-delivery to the river port, a significant part of overall transport costs. Sometimes, industries use a dedicated or quasi-dedicated terminal. The example of Bayer in the chemical industry is very relevant. Some shippers include the transhipment terminal directly on their industrial site. Finally, alliances between river carriers and rail carriers can provide new services of multimodal transport.

New Commercial Offers

The carriers' offers must provide 'door-to-door' transport services efficiently, reliably and at an attractive price by comparison with road transport. Longevity and financial stability entail aggregation and mini-mization of pre-and post-transhipment costs. Such conditions are difficult to meet. This requires the direct involvement of large customers and suppliers, and of course, public authorities. A 'door-to-door' service needs a coordinator to organize the transport chain. The price of this kind of service must of course be lower than that of road transport. Sufficient frequency and reliability, similar to road transport, are also key factors of success. Here, the transport time is less important than meeting delivery deadlines, especially when logistic flows of goods are based on the just-in-time model (for production and distribution). Additional services in relation to the road may also increase the use of IWT. These include, for instance, the ability to clear goods at the river port and not at the seaport. This saves time and money and allows the use of travel time by barge or the river terminal as a storage area to adjust the delivery time. Storage is often less expensive than at major sea terminals. A location near to a city favours proximity to the market and therefore to the end customer.

REFERENCES

ADEME (2008), *Transport combiné de marchandises: aides aux transporteurs et chargeurs*, Paris: ADEME.

Batsch, L. (1993), *La croissance des groupes industriels*, Paris: Economica.

Beamon, B.M. (1999), 'Designing the green supply chain', *Logistics Information Management*, **12**(4): 332–42.

Becker, D. (2003), *Le développement des implantations logistiques en France et ses enjeux pour les politiques d'aménagement*, Paris: Ministère de l'Équipement, des Transports, de l'Aménagement du Territoire, du Tourisme et de la Mer.

Bowersox, D.J. and D.J. Closs (1996), *Logistical Management: The Integrated Supply Chain Process*, New York: Macmillan.

Brundtland, G.H. (1987), *'Our Common Future', Report of the World Commission on Environment and Development (WCED)*, Geneva.

Christopher, M. (1999), 'Les enjeux d'une supply chain globale', *Logistique et Management*, **7**(1): 3–6.

Christopher, M. (2000), 'The agile supply chain: competing in volatile markets', *Industrial Marketing Management*, **29**(1): 37–44.

Christopher, M. (2011), *Logistics and Supply Chain Management: Creating Value-Adding Networks*, 4th edn, New York: Financial Times Prentice Hall.

CIADT (2003), *50 grands projets pour la France en Europe*, Paris: Matignon.

Clinton, S.R. and D.J. Closs (1997), 'Logistics strategy: does it exist?' *Journal of Business Logistics*, **18**(1): 19–44.

Dewberry, E. (1995), 'Ecodesign strategies', *Eco Design*, **4**(1): 32–3.

Dornier, Ph-P. and M. Fender (2007), *La logistique globale: enjeux, principes, exemples*, 2nd edn, Paris: Éditions d'Organisation.

Elram, L.M and M.C. Cooper (1990), 'Supply chain management, partnership, and the shipper–third party relationship', *International Journal of Logistics Management*, **1**(2): 1–10.

European Commission (2001), *White Paper – European Transport Policy for 2010: Time to Decide*, Luxembourg.

Fisher, M.L. (1997), 'What is the right supply chain for your product', *Harvard Business Review*, **75**(2): 105–16.

Guillaume, J-P. (2009), 'Des jouets acheminés par barges', *Supply Chain Magazine*, **36**: 42–3.

Gunasekaran, A. and E.W.T. Ngai (2004), 'Information systems in supply chain integration and management', *European Journal of Operational Research*, **159**: 269–95.

Hammer, M. and J. Champy (1993), *Re-engineering the Corporation: A Manifesto for Business Revolution*, New York: Harper Business Books.

Harland, C.M., R.C. Lamming, J. Zheng and T.E. Johnsen (2001), 'A taxonomy of supply networks', *Journal of Supply Chain Management*, **37**(4): 21–7.

Lambert, D.M., J.R. Stock and L.M. Ellram (2006), *Fundamentals of Logistics Management*, European edn, London: McGraw-Hill.

Maibach, M., C. Schreyer, D. Sutter, H.P. Van Essen, B.H. Boon, R. Smokers, A.Schroten, C. Doll, B. Pawlowska and M. Bak (2008), *Handbook on Estimation of External Costs in the Transport Sector*, Internalisation, Measures and Policies for All External Cost of Transport (IMPACT), CE Delft, The Netherlands.

Maxwell, D. and R. Van der Horst (2003), 'Developing sustainable products and services', *Journal of Cleaner Production*, 11: 883–95.

McKinnon, A. (2001), 'Integrated logistics strategies', in A.M. Brewer, K.J. Button and D.A. Hensher (eds), *Handbook of Logistics and Supply Chain Management*, London: Elsevier, pp. 157–70.

Melo, M.T., S. Nickelb and F. Saldanha-da-Gamad (2009), 'Facility location and supply chain management: a review', *European Journal of Operational Research*, **196**(2): 401–12.

Mentzer, J.T., W. DeWitt, J.S. Keebler, S. Min, N.W. Nix, C.D. Smith and Z.G. Zacharia (2001), 'Defining supply chain', *Journal of Business Logistics*, **22**(2): 1–25.

Mentzer, J.T., T.P. Stank and T.L. Esper (2008), 'Supply chain management and its relationship to logistics, marketing, production, and operations management', *Journal of Business Logistics*, **29**(1): 31–45.

Meunier, C. (1999), 'Infrastructures de transport et développement: l'apport de l'économie des réseaux', *Les Cahiers Scientifiques du Transport*, **36**: 69–85.

Miles, R.E., C.C. Snow and H.J. Coleman (1992), 'Managing 21st century network organizations', *Organizational Dynamics*, **20**(3): 5–20.

Ministère de l'Économie, de l'Industrie et de l'Emploi (MEIE) (2006), *Technologies clés 2010*, Paris.

Mortensen, O. and O.W. Lemoine (2008), 'Integration between manufacturers and third party logistics providers?' *International Journal of Operations and Production Management*, **28**(4): 331–59.

National Science and Technology Council (NSTC) (2001), *Vision 2050: An Integrated National Transportation System*.

Offner, J-M. (1993), 'Les effets structurants du transport: mythe politique, mystification scientifique', *L'espace géographique*, 3: 233–42.

Orsato, R. (2006), 'Competitive environmental strategies: when does it pay to be green?' *California Management Review*, **48**(2): 127–43.

Orsato, R. (2010), *Strategic Environmental Management: Competing for Natural Advantage*, London: Sage Publications.

Paché, G. (2006), 'Approches spatialisées des chaînes logistiques étendues: de quelles proximités parle-t-on?' *Les Cahiers Scientifiques du Transport*, **49**: 9–28.

Préfecture de Région Rhône-Alpes (2008), *Développement du fret fluvial sur l'axe Rhône-Saône: six grandes enseignes s'engagent à retirer 5.850 poids lourds de la route par an*, Communiqué de presse.

Savy, M. (1995), 'Morphologie et géographie des réseaux logistiques', in M. Savy and P. Veltz (eds), *Économie globale et réinvention du local*, DATAR, Paris: Editions de l'Aube, pp. 85–94.

Savy, M. (2007), *Le transport de marchandises*, Paris: Eyrolles.

Schmidheiny, S. (1992), *Changer de cap. Réconcilier le développement de l'entreprise et la protection de l'environnement*, Paris: Dunod.

Schmidt, J. (2005), 'Le développement d'une logistique en accord avec le développement durable', *Logistique et Management*, **13**(1): 31–6.

Stalk, G. and T. Hout (1992), *Vaincre le temps: reconcevoir l'entreprise pour un nouveau seuil de performance*, Paris: Dunod.

Sullivan, R. (2006), 'An introduction to corporate environmental management striving for sustainability', *Corporate Governance*, **6**(1): 96–8.

Ummenhofer, M. (1998), 'La logistique dans une perspective d'écologisation: vers l'éco-logistique intégrée', Doctoral thesis, Université de la Méditerranée, Aix-Marseille II.

Voies Navigables de France (VNF) (2008), *Transport fluvial: l'alternative durable*, Paris.

Voies Navigables de France (VNF) (2010), *Le trafic fluvial de 1999 à 2009*, Paris.

12. Integrating sustainability and technology innovation in logistics management

Matthias Klumpp, Sascha Bioly and Stephan Zelewski

INTRODUCTION

In logistics sustainability, concepts such as CO_2 footprint measurement, green logistics or empty transport reduction are developed and used in relative isolation from important future technology trends such as radio frequency identification (RFID), global positioning system (GPS) or process automation. This chapter seeks an integrated view of the relation between new technologies and sustainability in logistics – and therefore will contribute to the question of whether the two topics are friends (synergetic relation) or foes (competitive relation). This could help logistics service providers and other supply chain companies to understand the question of sustainability better and improve their future investment and business development planning.

In 2004 the logistics market in Germany alone amounted to €170 billion (Klaus and Kille, 2006: 43, 70, 80). Figures from the Fraunhofer Institute for 2007 report a logistics market volume of €205 billion in Germany and €900 billion in the whole of Europe in this sector (Fraunhofer Institut, 2008: 1), therefore this market is growing stronger than the general gross domestic product (GDP) and is ranked in third place after the automotive industry with €337 billion and the machinery industry with €219 billion (Klaus and Kille, 2008: 2).

The indicated 43 per cent transport costs in logistics amount to about €90 billion. The strategic position of logistics in any sustainability concept is also highlighted by the fact that, for example, for Germany the export rate additionally drives the logistics market. For Germany this export share of GDP has risen from 16 per cent in 1995 to 23 per cent in 2006 (Die Bundesregierung, 2008: 9). But logistics is also fuelled by the connected procurement inflows of goods, as most export products (and also

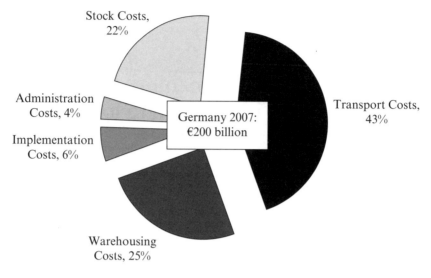

Source: Klaus and Kille (2008: 3)

Figure 12.1   Logistics market volume and functional segments

services) need the import of raw materials and modular parts, usually from Eastern Europe or Asia. Increasing competition is putting pressure on profit margins and disturbing future outlooks in different areas: 'trade companies are facing a logistics cost share 2008 of on average 15.9 per cent of total costs' (compared to industry companies: 7.0 per cent) (Straube and Pfohl, 2008: 4).

Important driving factors are rising fuel prices, road charges and ecological taxes as well as the standardization of logistics services (Göpfert and Hillbrand, 2005: 48). The German Logistics Association (BVL) is drawing the same bottom line and reporting: 'logistics costs [are] . . . basically driven by rising energy, fuel, transport and personnel costs' (Straube and Pfohl, 2008: 3). In this context BVL names the four most important trends for the future as globalization, sustainability, security and innovation (Straube and Pfohl, 2008: 1). This is highlighted by the following citation from the study:

> Additionally to globalization the logistics industry is harassed by increasing sustainability requirements, increasing security regulations and technology innovation expectations. For the companies asked in the study more and more security and stability questions are posed regarding their supply chains. Besides procurement and marketing risks there are mainly risks regarding natural

catastrophes, strikes, terror incidents or supply chain partners going out of business. Expected technology innovations are additionally said to influence their cost situation. And long-term views stress the importance of sustainable logistics concepts. (Straube and Pfohl, 2008: 1)

Moreover the improvement of service quality (flexibility, reliability, reactivity) as a logistics objective is ranked before cost reduction objectives – though most of the time, real cost transparency is not a given in logistics systems (Straube and Pfohl, 2008: 2).

In general, definitions of logistics refer to business administration terms and concepts. But nowadays, more and more, the reference framework for logistics concepts has to be enlarged towards ecological problems as logistics and transport processes contribute to energy consumption, pollution and noise emissions (Eickmann, 2002; Eisenkopf, 2006; Zelewski et al., 2008). Moreover there are specific risks in transporting dangerous goods, for example (Klaus and Krieger, 2004: 547). Interdisciplinary views of logistics concepts therefore have to integrate economical and ecological perspectives, as outlined below (Muchner, 1997: 41).

Logistics and innovation are a 'natural pairing' as shown by the basic optimization orientation of logistics by definition, as well as the following citation:

> There is no question if the logistics industry is innovative or not – but we lack the specific and empirically grounded knowledge about causes, success factors, hurdles and internal as well as external requirements for innovation competence, innovation actions and successful innovation management in industry and retail companies with core logistics functions (about 55 per cent share) as well as logistics service providers (about 45 per cent share). (Voß, 2008: 5)

Innovations are not restricted to single companies in supply chains: the current (2012) German central government is aiming at improving the leading frontline position of logistics in Germany. This is underpinned and detailed in the concept of the 'Master Plan Logistics Germany of 2008' and is rich with technology and innovation policies (Die Bundesregierung, 2008: 9).

Logistics is an interesting field for studying the interaction of sustainability and technology innovation as most companies, according to a PricewaterhouseCoopers (PWC) survey, invest in new technologies (new transport equipment, new handling equipment) in order to improve their ecological impact and at the same time their cost situation, as depicted in Figure 12.2.

At the same time the potential for improving sustainable solutions is still high in the logistics sector: for example, carbon footprint measurement is

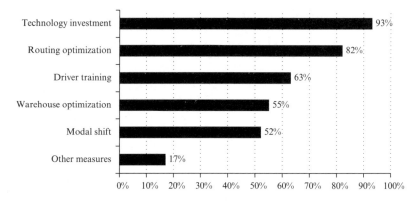

Source: PWC (2009: 9).

Figure 12.2   Sustainable investments in logistics

today expected by only 28 per cent of the customers of logistics service providers in Germany in the PWC study, but by 48 per cent of the customers of logistics service providers who act globally (PWC, 2009: 11). Therefore a growing interest and also impact potential for technology innovation in the context of green logistics solutions can be expected and should foster research interest in this field. Especially, the synergies or conflicts of logistics technology investments regarding sustainability should be discussed.

ECONOMIC PERSPECTIVE

In the logistics industry advertising and a common brand name may support a logistics service provider's success – however profits do not result from these external activities but from internal economic or business administration concepts and measurements. A study of CapGemini with 300 participating companies from the sector shows 'that stock keeping optimization or procurement price reduction has more impact on short-term profits than long-term supply chain management projects' (logistik-inside.de, 2009a).

Warehouse Concepts

Many companies for example follow warehouse centralization concepts, to be distinguished as vertical and horizontal centralization, with vertical defining the number of stock keeping levels and horizontal denominating the number of warehouses per level. Therefore in the horizontal dimension the number of warehouses defines the throughput in each single warehouse per stock keeping level, with centralization concepts enlarging this throughput significantly. Each warehouse has to handle a significantly higher number of goods in the same time period, enabling the company to use economies of scale in throughput handling as well as in transport delivery from this warehouse (but accepting higher transport costs caused by longer distances in the distribution area attached to one single warehouse).

In addition safety stock levels can be lowered. And most importantly for technology innovations, the number of affected locations when introducing new technologies, for example RFID, is diminished, reducing introduction costs significantly.

In this case innovation and sustainability may be foes, as a lower innovation barrier on the one hand results in significantly longer transport routes on the other hand (concept B). Assuming a constant throughput for concepts A and B (Figure 12.3) of 500 tonnes and an equal distribution of this throughput among warehouses, it is clear that concept A means 100 tonnes throughput per warehouse and a small distribution area.

In concept B one has to face longer transport distances but can operate with only two warehouses with 250 tonnes throughput each. Therefore a typical and modern logistics concept of warehouse centralization has the following effects on innovation and sustainability, showing that a final decision is not possible in general, but that each and every effect has to be measured in individual cases:

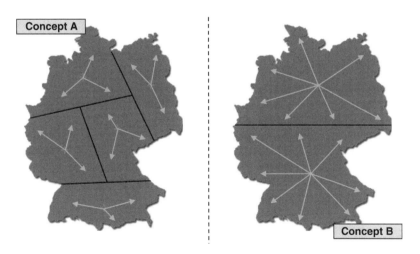

Figure 12.3   Stock keeping centralization (Germany)

- fixed cost reduction and higher economies of scale;
- lower innovation hurdles through lower numbers of logistics assets;
- potentially lower ecological impact (energy, pollution, noise) through larger transport volumes (bundling effects);
- potentially higher ecological impact through longer transport distances.

Stock keeping strategies usually depend on production and procurement strategies as well as distribution strategies, with higher lot sizes in production and procurement indicating more centralized warehouse structures than smaller lot sizes, and more points of origin (suppliers) and points of destination (customers). In addition first-rate communication, handling and transport technologies have to be implemented in order to operate successful central warehousing concepts (Delfmann, 2004: 520).

Logistics Service Levels

In order to control such concepts, key performance indicators (KPIs), for example stock keeping unit (SKU) availability or distribution service level, are measured in order to manage stock efficiently. Also this service level can be influenced by new technologies and has distinct impacts on sustainability: a low service level leads to missing stock costs (production or earnings failures) but allows for smaller warehouses and transport lot sizes and therefore a lower ecological impact. But a very high service level also implies high stock keeping costs as well as sustainability failures through (too) high stock levels and transport demands. Therefore an optimization in this area leads to possible synergies in both areas, economic and ecologic.

The logistics companies speak the same language: the ten most important supply chain management (SCM) projects today are driven 48 per cent by stock keeping measurements, 45 per cent by strategic supply chain measurements and 44 per cent by improvement of long-term planning (logistik-inside.de, 2009a).

This proves that technology innovations can contribute to both areas in terms of improving stock keeping as well as handling efficiency, leading to economic as well as ecological beneficial results. For example warehouse automation is greatly welcomed by all participants in logistics in order to reduce missing SKU in rolling in and rolling out processes in a warehouse.

Besides the physical effects of such new technologies the information technology (IT) area is highly important to such concepts, as an enabler as well as an instrument to further improve the planning and steering process in order to reach higher efficiency levels (ten Hompel, 2008: 16). It has to

Table 12.1   Results of insourcing vs outsourcing

	Strategic competitive advantage	Operative cost/quality advantage
Outsourcing	Economies of scale and external synergies by bundling similar logistics services for several customers	Reduction of factor costs by structural or opportunistic causes, integration of external logistics know-how
Insourcing	Economies of scale and internal synergies by bundling different services of one single customer	Profiteering of sales and profits of third companies by insourcing, supporting and extending internal know-how

Source: Klaus and Krieger (2004: 256).

be controlled accordingly that investment and change-over are lower than the savings, so that they become economically useful concepts.

Logistics Outsourcing

A further discussion point in logistics could be the ongoing improvement through insourcing or outsourcing concepts (Table 12.1): at first view this does not have much in common with technology development or innova tion. But in the details of logistics processes and outsourcing it can be recognized that most companies do implement new technologies as a core business advantage when tackling outsourcing.

These logistics descriptions can also be understood as technology changes. The economically efficient use of technologies follows similar routes as described for logistics outsourcing above. Moreover, outsourcing brings in new technologies ('external know-how') which highlights the important role of producers and logistics service providers as change agents in logistics outsourcing projects. This could lead to operative cost and quality advantages – sometimes also to improved resource efficiency in terms of energy consumption for example. This depends to a great extent on the right point in time for investing in such new technologies and now has an increased importance due to global supply chains. Efficient information use and transmission is a very important subtask in order to steer stock levels and transport flows (Klaus and Krieger, 2004: 455).

This should lead to the reduction of out-of-stock situations, requiring modern information architectures and systems in order to avoid negative effects on stocks and transport flows (bullwhip effect). This effect is also called the 'Forrester', 'whiplash' or 'whipsaw' effect, and describes the empirical measured effect of rising lot and stock sizes along a supply chain due to lack of information. As described in the stock keeping concepts, this could lead to inefficiencies economically as well as ecologically due to more transport than necessary and larger warehouse sizes with increased stock levels (Klaus and Krieger, 2004: 455).

In addition to the temporal aspect of the effect due to globalization the resulting fragmentation of the value chain is affiliated to a considerable local aspect, which induces traffic. That means several countries may be affected and therefore consuming a lot of energy and bearing high costs. One measure against these effects is called baton passing: this describes overlapping frontiers between two activities, meaning that one does not have to be finished completely in order to start the second activity, for example transporting the goods to the sales house (Klaus and Krieger, 2004: 36).

Another option is the so-called vendor-managed inventory: in this case the supplier receives constant information about the changes in stock and transport status of all goods in the supply chain. According to this information the supplier steers the supply of goods in the supply chain and can therefore possibly reduce stock levels and transport flows (ten Hompel, 2008: 237).

All these concepts rely on best possible technologies for information recognition (automated) and analysis by computers. The better and faster this can be achieved through technology (Barcode, IT, RFID, GPS), the better can steering impulses and therefore efficiencies be achieved.

ECOLOGICAL PERSPECTIVE

As early as December 1993 the phrase was coined that 'environmental issues are one of the most important of our time' (Ventzke, 1993: Preface). The general importance of sustainability concepts has more likely risen than fallen in the years since then. The public consciousness in terms of such issues is mainly aroused by major incidents such as Seveso (1976), Chernobyl (1986), *Exxon Valdez* (1989) or Fukushima (2011). These examples are more or less tragic incidents which could have been avoided in each and every case – though there is a general overall statistical probability for such events to happen.

Logistics has to face another significant ecological threat: the effects and causes of 'creeping' emissions in the context of logistics and transport processes. For example the pollutants CO_2 and NO_x are emitted in road transport and travel activities. Therein CO_2 accounts for roughly 9–26 per cent of the global greenhouse effect and NO_x support ozone layer destruction (Kiehl and Trenberth, 1997: 197). In a long-term perspective the greenhouse and ozone displacement effects are interdependent concepts – though they are basically two different phenomena in terms of causes as well as reduction concepts.

For the ozone layer, hazardous chlorofluorocarbons (CFCs) and also ozone (O_3) itself belong to the greenhouse gases. They are nevertheless not causing huge negative effects or harm because of their small volume. The Max Planck Institute for Meteorology in Hamburg, despite this, warns that an increasing greenhouse effect may accelerate the ozone layer destruction (Max Planck Institute, 2009). Since 1994 the ozone hole has been closing up – leaving scientists in doubt about the forecast of near recompletion. The National Aeronautics and Space Administration (NASA), for example, estimates that about the year 2068 will bring a near-original status of the ozone layer (NASA, 2009).

This will require totally new technology concepts in order to reduce such pollution. This can happen by way of different concepts based in different technologies, which can be described as follows:

● Reduction of direct emissions by road transport, for example by more efficient motors, energy recycling systems, hybrid driving systems and others.
● An optimization of equipment used by implementing state-of-the-art information technologies in order to reduce, for example, empty transport legs in different transport systems such as road, rail, sea or air.
● A contribution to a reduction of waste could be reusable packaging. This would bring a closed-loop system to many supply chains from the logistics point of view. Optimized packaging could help to reach more economic and also ecological transport processes and therefore poses a perfect synergy potential for these areas – though many of those could not necessarily be termed 'high-tech' (Ullrich, 1996: 3).
● The secure handling of goods transport, especially in the areas of recycling and dangerous goods, is an important and environmentally crucial task in logistics. The relevant German legislation (GGBefG) defines not only the transport but also the handling processes (loading, unloading, storing) as relevant under this regulation. Also in this area new technologies in terms of avoiding, handling and storing such waste and dangerous goods can improve the sustainability concept in logistics. For example environmental 'hot spots' are the final storage points for radioactive material in Asse II (Wolfenbüttel) and Gorleben in Germany.

- A reduction of traffic and pollution especially in inner-city logistics by cooperative city logistics concepts can be supported by matching innovative technology concepts. This requires the cooperation of several logistics companies and aims at efficient transport flows in highly populated areas (inner cities) with their traffic restrictions (Klaus and Krieger, 2004: 85).

As green logistics concepts gain importance and publicity, many logistics service providers and also industry companies offer such concepts and projects regarding sustainability: for example Procter & Gamble promises to raise the share of rail transport from 10 per cent to 30 per cent until 2015 in order to reduce road traffic with all its emissions and negative external effects. The products of this company travel 200 million kilometres in Western Europe annually. The announced modal change towards rail transportation would potentially save up to 67 500 tonnes of CO_2 per annum (logistik-inside.de, 2009b).

The German logistics service provider DHL/Deutsche Post is offering business customers the opportunity to post their surface mail 'climate change neutrally' as 'Go Green' products – with a promised complete analysis of all emitted CO_2 amounts during the transport and the later remission by, for example, tree planting in order to offset the emitted CO_2 (logistik-inside.de, 2009b).

Moreover many concepts are implemented today in Germany and the European Union in order to reduce the emissions of respirable particles by trucks and cars in order to improve quality of life and also public health in cities. Studies show that such reductions can increase quality of life and also life expectation significantly, by 15 per cent. 'A reduction of respirable dust emissions of ten microgram per cubic meter air has increased the life expectancy by 0.6 years on average' according to the epidemiologist C.A. Pope of Brigham-Young-University in Provo (Utah), writing in the *New England Journal of Medicine* (Frankfurter Rundschau online, 2009).

Further topics for the interaction of technology and innovation in sustainable concepts are, for example:

- land use;
- energy consumption;
- air pollution;
- resource consumption;
- residues;
- water consumption and water pollution;
- as well as other ecological effects.

ELECTRIC MOBILITY

Here, another aspect is examined which links innovation with sustainability: electric mobility. It is certainly a technological development and may be examined to be 'friend or foe' to sustainable logistics in this context. Currently it can be assumed that electric cars will be part of our daily lives before electric trucks. This is the reason why cars are described here first.

Passenger traffic is of outstanding importance for our individual mobility. This fact is illustrated by the data for the development of car use not only in Germany, but worldwide. Sales of the automotive industry in Germany in 2009 – the year of financial and economic crisis – were €278 billion (2008: €345 billion) and this is therefore, in terms of turnover, the most important sector in Germany. In second place is engineering with €175 billion (2008: €225 billion) (Statistisches Bundesamt, 2010: 370–375).

Undisputedly, oil is currently the most important energy source; for all established transport systems it has great importance and, at least in the short and medium term, oil will retain this importance.

The combustion of fossil fuels such as lignite or hard coal, natural gas and crude oil, due to the so-called carbon cycle, inevitably causes the emission of carbon dioxide (CO_2). According to the International Energy Agency (IEA) in 2009, nearly 81 per cent of global energy needs were covered by fossil fuel sources (International Energy Agency, 2011: 6).

Ever since the decision of the Additional Protocol of the United Nations Framework Convention on Climate Change (UNFCCC) – better known as the Kyoto Protocol – on 11 December 1997 in Kyoto (Japan), the reduction of greenhouse gas emissions has been a target of many nations. Currently 193 countries have ratified this agreement (UNFCCC, 2012). On 13 December 2011 the Süddeutsche Zeitung reported that Canada had officially withdrawn from the Kyoto Protocol, and attributed this to the fact that the emissions standards are not accepted by the United States (USA), as well as some emerging and developing countries (such as China and India), and only 37 industrialized nations in total have respected its specifications. Earlier, Japan and Russia already announced their disagreement to an extension of the future requirements (the initial commitment period of Kyoto Protocol was 2008–12), as long as the biggest emitters of greenhouse gases (the USA and China) would not be included (Süddeutsche Zeitung, 2011).

The desire to reduce greenhouse gas emissions, due to the use of fossil fuels, especially oil, forces the search for alternative concepts.

Starting Position

Through technological advances, recorded world oil reserves have been increased in recent years. But oil, as any fossil fuel, is not an unlimited resource. The increasing scarcity of oil due to its finite nature is exacerbated by other factors, including the security problems and political turmoil in some oil-producing countries (about 62 per cent of the oil reserves are attributable to the countries of the Middle East) and rising oil prices due to growing demand (in particular, due to economic growth in the BRICS countries of Brazil, Russia, India, China and South Africa), and thus leads to more speculative markets. Therefore it seems sensible to reduce the oil dependence of mobility in the future, or at best even eliminate the addiction.

Electric mobility can lead to a way out of this situation. Vehicles require energy to move. Combustion engines use chemical energy stored in the fuel by combustion – with associated waste products such as CO_2. For an electric vehicle, which is completely or partially electrically driven, the electrical energy can be kept in vehicle batteries; these batteries, after use (discharge), have to be recharged. Looking at the current range of electric vehicles, it is found that these do not come close to the reach of conventional vehicles. This is due to the different energy storage capacity (gravimetric energy density) of petrol, diesel and advanced batteries.

Climate Change

Climate change could have dramatic consequences for humanity. The first harbingers are already visible: melting glaciers and polar ice, rising sea levels, extreme weather events and damage to the ecosystem. Although it is not clear how much of these developments are due to man, the greenhouse effect which arises due to greenhouse gases emitted by humans is called the anthropogenic greenhouse effect, and this effect is greater than zero. It impacts on climate in addition to the natural greenhouse effect (Schönwiese, 2003: 340). This means that mankind has caused on the one hand the increased concentration of natural greenhouse gases (e.g. carbon dioxide, methane, nitrous oxide), and on the other has created new synthetic greenhouse gases, such as CFCs and chlorfluoromethanes (CFM) (Table 12.2).

The IPCC (Intergovernmental Panel on Climate Change), headquartered in Geneva, was founded in 1988 by the United Nations Environment Programme (UNEP) and the World Meteorological Organization (WMO). The committee works on the consequences and risks of climate change and explores how they can mitigate or adapt to it.

Table 12.2   Anthropogenic greenhouse gases emitted

Kind of gas	Anthropogenic development process	Trend 1800–2000*	Contribution to the anth. greenhouse effect (%)	GWP in CO_2e
Carbon dioxide (CO_2)	● Use of fossil energy ● Deforestation	280 ppm → 370 ppm	61	1
Methane (CH_4)	● Use of fossil energy ● Livestock ● Rice cultivation	0.28 ppm → 0.31 ppm	15	25
Nitrous oxide (N_2O)	● Especially soil (fertilizer)	Not specified	4	298
CFC gases	● Artificial creation in propellant (spray cans) ● Coolant	0 ppb → 0.5 ppb	11	< 14.400
Sulfur hexafluoride (SF_6)	● Shielding gas in the industrial production of magnesium	0 ppb → 40 ppb	4	22.800

Note: * Estimates; ppm ≙ parts per million (10^{-6}); ppb ≙ parts per billion (10^{-9})

Source: Schönwiese (2003: 337–49).

Causes of climate change according to the fourth IPCC report (IPCC, 2007: 35–41) are as follows:

● The carbon dioxide content of the air has increased since 1750 by 35 per cent from 280 ppm to 379 ppm in 2005. The growth rate of the last ten years has been the biggest in 50 years. The present value is the largest in the last 650 000 years. 78 per cent of the increase is due to the use of fossil fuels and 22 per cent to land use changes (by deforestation).

● Other relevant greenhouse gases have increased, such as methane and nitrous oxide: their concentrations have increased since 1750 by 148 per cent and 18 per cent, respectively, together about half as much as the CO_2 increase.

Climate change and the limited availability of fossil fuels have caused the plan to reduce greenhouse gas emissions and make efficient use of natural resources to became a central issue of our time.

In this context it is important not only to talk about the costs that result from climate change, but also to demonstrate the benefits of more efficient use of energy resources. Meanwhile, the measures developed to reduce anthropogenic climate change additionally ensure future prosperity (Ziegler, 2011). The integration of infrastructure into the day-to-day lifestyle is rarely called into question. Hybrid vehicles and a green lifestyle do not fit into a suburban lifestyle. It is not environmentally friendly or sustainable if basic services such as food can only be obtained by making a long journey (Schützenmeister, 2010). The question of sustainability is discussed more and more, especially in logistics (Halldorsson et al., 2009; Klaus, 2009; Klumpp, 2009).

However, energy and resource consumption must be reduced by reducing the emissions of pollutants such as CO_2 (Browne et al., 2001; Zelewski et al., 2008; Al-Mansi et al., 2008). Leaps in efficiency are no longer expected by increasing the efficiency of motors. Therefore, new concepts, new modes of transport, new propulsion concepts have to be developed (Stan, 2005; Klaus and Krieger, 2004).

Three More Reasons for the Strengthening of Electric Mobility

Firstly, demographic change. From a European point of view demographic change is usually understood as an ageing population in combination with overall declining population numbers. This view of demographic change is not regarded as a globally uniform one. In particular, older people assume that their personal mobility will tend to decrease with increasing age. Numbers of older people are higher than several years ago, and the question is in what environment they can live for as long as possible without needing help from others, and how they will be able to satisfy their social needs (cinema, theatre, etc.). In addition, younger people avoid sparsely populated regions, and the challenges these pose for themselves and their children. This results in the second point.

Secondly, urbanization, seen as migration. Push and pull factors can be distinguished. Urban areas are traditionally places with more attractive opportunities for people's private as well as professional lives, and so to live in these areas is many people's goal (pull). These regions tend to offer more (quantitative) and better (qualitative) job opportunities. Rural regions are often limited in economic and recreational opportunities (which can be seen as push factors). This aspect is influenced by the above described, demographically induced change.

Thirdly, protection of resources. The conservation of resources is seen as a challenge. Extensive urban sprawl results in a larger space requirement, and the resulting lower population density makes it difficult to assure widespread access to mass transportation systems such as subways and other modes of public transport. The more compact urban structures associated with shorter distances allow and encourage, among other things due to the higher population density, the absence of individual mobility by car, without (or with a moderate but acceptable) loss of comfort, in favour of public transport.

With increasing urbanization the distances could be shortened, which favours the use of electric vehicles under the present conditions with respect to range and charging times. Overall, the trends in urban development as well as in politics point in the same direction: modern logistics concepts, with the electrification of cars as a starting point (UN World Urbanization Prospects, 2008; Vogelpohl, 2011: 234; Brake, 2011: 299; Davis, 2007: 85–9; Swiaczny, 2008: 449–56; Sassen, 1996; Arning and Reuther, 2008; Brake, 1993).

Current Propagation of Electric Vehicles

Figures from the Federal Motor Transport Authority indicate 42.3 million passenger cars in January 2011 in Germany, which are 90.2 per cent in private hands; and alternative technologies are accepted only reluctantly (Kraftfahrtbundesamt, 2011). So far (2012) only about 40000 cars with hybrid/electrical engines have been registered. On 3 May 2010 a statement was given by the federal government: 'up to 2020 . . . at least 1 million electric vehicles should be on German roads' (Federal Government of Germany, 2010). This represents about 2 per cent of all vehicles (cars only) and poses major challenges.

For conventional internal combustion engines, the tank capacity is usually dimensioned for a minimum range greater than 500 km to the next refill. For electric vehicles currently maximum ranges are expected at around 150 km per battery charge. Assuming that the batteries cannot be completely discharged, it can be assumed that the charging intervals have to be at least 4–5 times shorter than for combustion engines. To make matters worse, the duration of this process is much longer. At petrol stations, petrol and diesel can be refuelled at 35 litres per minute, and the whole filling process is completed in under two minutes. This gives a theoretical duration of 24 seconds for a 100 km range. For electric vehicles, six hours (at 230 V) for 150 km has to be expected, the equivalent of four hours per 100 km; or alternatively 30 minutes for an 80 per cent load at 400 V, resulting in 25 minutes per 100 km range (all data are from manu-

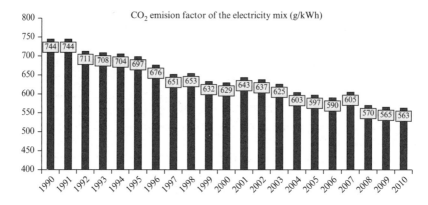

Source: Federal Environmental Agency (2011: 2).

Figure 12.4 CO_2 emissions caused by the electricity mix in Germany

facturers' information for the Citroën C-Zero, Mitsubishi i-MiEV, Nissan Leaf, Peugeot iOn and Renault Kangoo Z.E.).

Petrol- and diesel-powered cars can be prepared 600 times faster for 100 km distances; the related area and time requirements for electric vehicles with a petrol station network as it is now available in Europe is unthinkable in the sense of sustainability.

Environmental Balance of Electric Vehicles

How much CO_2 emissions are actually caused by electric vehicles? Locally, while driving, the often-used slogan 'zero emissions' is correct: 0 grams of CO_2 per kilometre. But it is necessary to consider carbon dioxide which is the result of the production of the charged electric energy. Thus, the overall balance depends on the so-called energy mix (Figure 12.4) which includes the source of the electricity (Federal Environmental Agency, 2011: 2–6).

In this figure the phase-out of nuclear power adopted by the Federal Government in 2011 and its consequences are not yet included. The current 22.5 per cent share of nuclear energy, which affects the CO_2 balance positively, would need to be fully substituted by low-CO_2 energy sources to keep the balance of electric vehicles at the same level.

Eco-electricity in Germany causes relatively low CO_2 emissions: about 40 grams of CO_2 per kWh of electricity. The highest CO_2 emissions from electricity generation are coal-fired and brown coal-fired power

Table 12.3   Types of power plants in CO_2 comparison

Type of power plant	CO_2 emissions (g/kWh)
Wind energy offshore	23
Wind energy onshore	24
Solar power [import from Spain]	27
Nuclear power plant	32
Hydroelectric power station	40
Multicrystalline solar cell	101
Natural gas combined power/heating plant	148
Natural gas power plant	428
Coal-fired power/heating plant	622
Brown coal power station	729
Coal-fired power plant	949
Lignite power plant	1.153

Source: Herminghaus (2012).

plants with 949 g CO_2 per kWh and 1153 g CO_2 per kWh, respectively (Herminghaus, 2012). Nuclear power plants also cause relatively low CO_2 emissions from electricity generation: 32 g of CO_2 per kWh of electricity. Since the problem of the treatment and final storage of nuclear waste is not yet clear, the environmental balance of nuclear power remains controversial despite the CO_2 emissions. Table 12.3 presents the available types of power plants in Germany and their average CO_2 emissions per kWh.

The so-called renewable energies in general are often mistakenly viewed as CO_2-free. But 'the issue of environmental sustainability depends very heavily on the energy mix', and Bosch chief executive officer (CEO) Franz Fehrenbach holds under these assumptions a conventionally powered 3 litre 'clean diesel' car to be more environmentally friendly than pure electric cars (Wagenhofer, 2011). This is especially true for markets such as China, whose electricity is derived mainly from coal, which creates a lot more carbon dioxide emissions. An electric drive can only make ecological sense when the cars are designed to be as lightweight as possible and to consume as little energy as possible. The equipment and streamlined bodies of these cars should be produced completely with renewable energy sources (hydro, wind, solar, biomass). In Germany about 580 grams of CO_2 per kilowatt hour of electricity are emitted by a high proportion of coal-fired power plants (57 per cent). In France, this figure is less than 100 g/kWh; in China it is more than 1000 g/kWh; in Austria the value is 155 g/kWh in 2008 – the environmental performance of electric vehicles is therefore largely determined by the current mix of energy used in electricity generation (Wagenhofer, 2011).

LOGISTICS AND SUSTAINABILITY IN BUSINESS CYCLES

The global financial crisis has had two different impacts on technology investments: on the one hand, technologies and investments are being introduced slower than planned; on the other hand there are also companies increasing their investments (antonym movement): according to a PricewaterhouseCoopers (PWC) study 42 per cent of logistics companies are postponing technology and sustainability investments due to the crisis, whereas 33 per cent are increasing their investments in order to cut costs (PWC, 2009: 12).

In sea shipping logistics similar investment reductions are also reported. For example Hapag-Lloyd is reacting to the economic crisis with cost reductions worth about €400 million (logistik-inside.de, 2009c). 'Every investment possible is postponed' says the CEO Michael Behrendt (Hamburger Abendblatt, 2009). The Hamburg Port Authority HHLA is planning to cut investment projects due to work and turnover shortages (logistik-inside.de, 2009d).

But in the logistics sector there are also companies with increasing investment plans: for example the German company Voigt Logistik is not postponing investments in new warehouses and trucks. But road and bridge building projects are profiting especially from the business crisis. According to PWC this is a major opportunity for the German traffic infrastructure – the German government plans to invest €14 billion in this sector (PWC, 2009). In a CapGemini study two out of three supply chain managers said that the crisis is the most important factor in their decisions. As the second most important factor, customer requirements were named, and the third was referred to as the quest for sustainability (logistik-inside.de, 2009a). In general it can be said that technology enhancements are necessary from an economic as well as an ecological perspective – and this is independent from business cycles.

INTEGRATED DEVELOPMENT PERSPECTIVE

An integrated perspective with a joint analysis of technology innovation and sustainability improvements should help, especially in the logistics sector, to improve technologies and also to provide economic cost reductions in order to enhance ecological efficiency at the same time.

The analysis scorecard in Figure 12.5 may contribute as a first suggestion to this concept of improvement in managing logistics innovation and sustainability in logistics (Klumpp and Ostertag, 2008; Jasper and Klumpp, 2008).

Future contributions may improve this first draft in terms of empirical piloting and testing as well as concept amendments.

CONCLUSION

As shown by this contribution to research as well as for practitioners, the question of the role of technology innovation and sustainability in logistics is not an easy one to address. Therefore more emphasis has to be put on further research and empirical evaluation. But even so some major hypotheses can be put forward in the direction of practical logistics implementation and projects.

First of all, the objectives of both technology innovation and sustain-

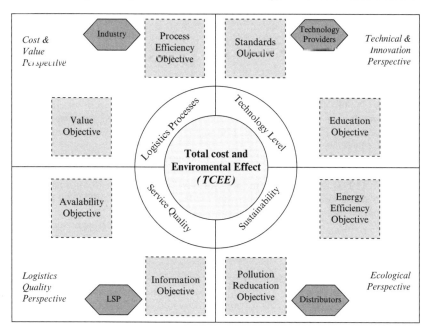

Figure 12.5   Technology Innovation and Sustainability Assessment in Logistics (TISAL)

ability should be included in any planning and concepts outlining activities in logistics by all partners in supply chains.

Secondly, the awareness has to be enhanced that there are some crucial concept decisions, as for cxample shown for warehousing structures, which define the interaction of technology innovation and sustainability on a long-term basis. Therefore such decisions should be reviewed more comprehensively and in depth, including the two named objectives as decision criteria.

Thirdly, a 'positive hunt' for best practice examples should be established in companies searching for and communicating positive examples for concepts integrating technology innovation and sustainability in logistics and therefore providing synergies between these two objectives.

Fourthly, in most cases technology innovation and sustainability should be seen as 'friends' with a synergetic relationship, especially with very long-term investment and business perspectives. Therefore the assumption of a synergetic relationship should be used as a default value in order to avoid misleading investments and conform with long-term political and social interests (the stakeholder concept).

This can enhance the chances to find technology and sustainability on the same side as friends instead of foes – which would benefit all partners in modern supply chains as well as the overall society.

REFERENCES

Al-Mansi, A., H. Aldarrat, F. Rhoma, A. Goudz, M. Lorenz and B. Noche (2008), 'Design of an environmental supply chain network: a biosolid waste case study', in W. Kersten, T. Blecker and H. Flämnig (eds), *Global Logistics Management*, Berlin, pp. 87–98.

Arning, J. and I. Reuther (eds) (2008), *Regiopolen. Die kleinen Großstädte in Zeiten der Globalisierung*, Berlin.

Brake, K. (1993), 'Die räumliche Struktur der Dienstleistungsökonomie – oder: warum gibt es keine Dezentralisierung?', in H. Häußermann and W. Siebel (eds), *Strukturen einer Metropole*, Frankfurt am Main, pp. 91–107.

Brake, K. (2011), '"Reurbanisierung" – Globalisierung und neuartige Inwertsetzung städtischer Strukturen "europäischen" Typs', in O. Frey and F. Koch (eds), *Die Zukunft der europäischen Stadt*, pp. 299–323.

Browne, F.E., P.D. Cousins, R.C. Lamming and A.C. Faruk (2001), 'Intelligent logistics – more sustainable freight traffic', in DVWG (ed.), *Traffic and Transport 2030 – Visions, Concepts, Technologies*, Berlin, pp. 69–79.

Davis, M. (2007), *Planet of Slums*, London, UK and New York, USA.

Delfmann, W. (2004), 'Logistische Zentralisierung', in P. Klaus and W. Krieger (eds), *Gabler Lexikon Logistik*, Wiesbaden, pp. 519–20.

Die Bundesregierung (2008), *Masterplan Güterverkehr und Logistik*, Berlin.

Eickmann, C. (2002), 'Ermittlung der CO_2-Emissionen, Methoden zum

Vergleich im Straßen-und Schienenverkehr', in Eurailpress (ed.), *EI – Der Eisenbahningenieur*, **53**(9): 116–22.

Eisenkopf, A. (2006), 'Ökonomische Instrumente für einen umweltvertraglichen Verkehr-Machbarkeit und Wirksamkeit', in Institut für Technikfolgenabschätzung und Systemanalyse (ITAS) (ed.), *TECHNIKFOLGENABSCHÄTZUNG – Theorie und Praxis*, **15**(3): 21–30.

Fraunhofer Institut (ed.) (2008), Presseinformation, Fraunhofer Institut Integrierte Schaltungen – Arbeitsgruppe für Technologien der Logistik-Dienstleistungswirtschaft ATL, Nürnberg.

Göpfert, I. and T. Hillbrand (2005), 'Innovationsmanagement für Logistikunternehmen', in Wolf-Kluthausen, H. (ed.), *Jahrbuch Logistik 2005*, Korschenbroich, pp. 48–53.

Halldorsson, A., H. Kotzab and T. Skjott-Larsen (2009), 'Supply chain management on the crossroad to sustainability: a blessing or a curse?' *Logistics Research*, **1**(2): 83–94.

International Energy Agency (IEA) (2011), *Key World Energy STATISTICS*, Paris.

IPCC (2007), Solomon, S., D. Qin, M. Manning, Z. Chen, M. Marquis, K.B. Averyt, M. Tignor and H.L. Miller (eds), 'Contribution of Working Group I to the Fourth Assessment Report of the Intergovernmental Panel on Climate Change, Synthesis Report', Cambridge, UK and New York, USA.

Jasper, A. and M. Klumpp (2008), 'Success factors for retail logistics in an e-commerce-environment', *Sinoeuropean Engineering Research Journal*, **1**: 63–8.

Kiehl, J. T. and K. Trenberth (1997), 'Earth's annual global mean energy budget', *Bulletin of the American Meteorological Society*, **78**: 197–208.

Klaus, P. (2009), 'Looking out for the next generation research questions in logistics', *Logistics Research*, **1**(3): 129–30.

Klaus, P. and C. Kille (2008), '*Zusammenfassung zu: Die Top 100 der Logistik. Marktgrößen, Marktsegmente und Marktführer in der Logistikdienstleistungswirtschaft*, Ausgabe 2008/2009, Hamburg.

Klaus, P. and W. Krieger (2004), 'Gabler Lexikon Logistik-Management logistischer Netzwerke und Flüsse', Wiesbaden: Gabler.

Klumpp, M. (2009), 'Logistiktrends und Logistikausbildung 2020', ild Schriftenreihe Logistikforschung No. 6, 12/2009, GER Essen.

Klumpp, M. and M. Ostertag (2008), 'Quality management impact on logistics networks measured by supply chain performance indicators', in W. Kersten, T. Blecker and H. Flaming (eds), *Global Logistics Management – Sustainability, Quality, Risks*, Berlin, pp. 129–48.

Muchner, C. (1997), 'Logistik im Spannungsfeld zwischen Ökonomie und Ökologie', in K. Inderfurth, M. Schenk and D. Ziems (eds), *Logistik auf Umweltkurs: Chancen und Herausforderungen*, Magdeburg, pp. 41–56.

PricewaterhouseCoopers (PWC) (ed.) (2009), *Land unter für den Klimaschutz? Die Transport-und Logistikbranche im Fokus*, Hamburg.

Sassen, S. (1996), 'Metropolen des Weltmarktes. Die neue Rolle der Global Cities', Frankfurt.

Schönwiese, C.-D. (2003), *Klimatologie*, Stuttgart.

Schützenmeister, F. (2010), 'Hybrid oder autofrei? – Klimawandel und Lebensstile', in M. Voss (ed.), *Der Klimawandel: VS Verlag für Sozialwissenschaften*, Wiesbaden: Springer, pp. 267–81.

Stan, C. (2005), *Alternative Antriebe für Automobile – Hybridsysteme, Brennstoffzellen, alternative Energieträger*, Berlin and Heidelberg: Springer.

Statistisches Bundesamt (2010), *Statistisches Jahrbuch 2010 für die Bundesrepublik Deutschland mit 'Internationalen Übersichten'*, Wiesbaden.

Statistisches Bundesamt (2011), *Statistisches Jahrbuch 2011 für die Bundesrepublik Deutschland mit 'Internationalen Übersichten'*, Wiesbaden.

Straube, F. and H. Pfohl (2008), *Trends und Strategien in der Logistik, Die Kernaussagen*, Bundesvereinigung Logistik (BVL), Berlin.

Swiaczny, F. (2008), 'Urbanisierung und Entwicklung 2008', *Zeitschrift für Bevölkerungswissenschaft*, **33**: 449–60.

ten Hompel, M. (2008), *Taschenlexikon Logistik. Abkürzungen, Definitionen und Erläuterungen der wichtigsten Begriffe aus Materialfluss und Logistik*, Berlin and Heidelberg, Germany and New York, USA.

Ullrich, R. (1996), *Mehrweg-Transportverpackungen*, Hamburg.

UN World Urbanization Prospects (2008), *World Urbanization Prospects. Executive Summary*, New York.

Ventzke, R. (1993), *Umweltorientierte Produktionsplanung: Ein analytischer Ansatz zur Berücksichtigung von Restriktionen in der Produktions- und Kostentheorie*, Münster.

Vogelpohl, A. (2011), *Städte und die beginnende Urbanisierung – Henri Lefebvre in der aktuellen Stadtforschung*, scientific contribution published online in Springer-Verlag.

Zelewski, S., A. Saur and M. Klumpp (2008), 'Co-operative rail cargo transport effects', *MAEKAS* Projektbericht, No. 2, Institute for Production und Industrial Information Management, University of Duisburg-Essen, Essen.

Ziegler, M. (2011), *Klimawandel und Energieeffizienz Kosten und Nutzen für die Wirtschaft*, Berlin and Heidelberg: Springer.

Web References

Federal Environmental Agency (Umweltbundesamt) (2011), 'Entwicklung der spezifischen Kohlendioxid-Emissionen des deutschen Strommix 1990-2009 und erste Schätzung 2010 im Vergleich zu CO_2-Emissionen der Stromerzeugung', available at http://www.umweltbundesamt.de /energie/archiv/ co2-strommix.pdf (accessed 1 January 2012).

Federal Government of Germany (2010), 'Etablierung der Nationalen Plattform Elektromobilität – Gemeinsame Erklärung von Bundesregierung und deutscher Industrie', available at http://www. bundesregierung.de/Content/DE/ Artikel/2010/05/2010-05-03-elektromobilitaet-erklaerung.html (accessed 26 January 2012).

Frankfurter Rundschau online (2009), 'US study: less particulate matter increases life expectancy', available at http://www.fr-online.de/wissenschaft/us-studie-weniger-feinstaub-erhoeht-lebenserwartung,1472788,3251476.html (accessed 19 March 2012).

Herminghaus, H. (2012), 'CO_2-Vergleich bei der Stromerzeugung in Deutschland', available at http://www.co2-emissionen-vergleichen.de/Stromerzeugung/CO2-Vergleich-Stromerzeugung.html (accessed 9 February 2012).

Kraftfahrtbundesamt (2011), 'Jahresbilanz des Fahrzeugbestandes am 1. Januar 2011', available at http://www.kba.de/cln_032/nn_125264/DE/Statistik/

Fahrzeuge/Bestand/   bestand__node.html?__nnn=true#rechts   (accessed   24 January 2012).

logistik-inside.de (2009a), 'Study: short-term storage optimization has priority', available at http://www.verkehrsrundschau.de/cms/829264 (accessed 19 March 2012).

logistik-inside.de (2009b), 'Procter & Gamble reinforce Green Logistics', available at http://www.verkehrsrundschau.de/cms/827984 (accessed 19 March 2012).

logistik-inside.de (2009c), 'Hapag-Lloyd expresses costs by 400 million €', available at http://www.verkehrsrundschau.de/cms/826856 (accessed 19 March 2012).

logistik-inside.de (2009d), 'HHLA slows investments – short-time work planned', available at http://www.verkehrsrundschau.de/cms/828626 (accessed 19 March 2012).

Max Planck Institute (2009), 'From the past to the future: new climate simulations for science and society', available at http://www.mpimet.mpg.de (accessed 14 March 2012).

NASA (2009), available at http://www.earthobservatory.nasa.gov/ (accessed 17 March 2012).

Süddeutsche Zeitung (2011), 'Canada escapes pact on climate change from Kyoto Protocol officially', available at http://www.sueddeutsche.de/politik/kli maschutzabkommen-kanada-steigt-offiziell-aus-kyoto-protokoll-aus-1.1233232 (accessed 22 January 2012).

UNFCCC (2012), 'Status of ratification of the Kyoto Protocol', available at http:// unfccc.int/kyoto_ protocol/status_of_ratification/items/2613.php (accessed 26 January 2012).

Wagenhofer, P. (2011), 'E-Autos brauchen sauberen Strom', available at http:// news.orf.at/stories /2059019/2059018/ (accessed 9 February 2012).

13. Sustainable development, a new source of inspiration for marketing innovation? Focus on five major trends and one innovative project in customer relationship marketing

Gaël Le Boulch and Rémy Oudghiri

INTRODUCTION

Even though US (United States) policies adopted by the Obama administration have helped stimulate international momentum for sustainable development, the number of companies that are clearly positioned in this fast-growing market remains limited. Is this out of timidity? Fear of diving into a realm that will be difficult to control or overly regulated? Is there incompatibility with certain sectors (i.e. luxury products) and their underlying theme that is contrary to the pleasure principle? Despite this hesitation, a recent study conducted by LinkedIn[1] (2009) pointed out the extent to which individuals are attentive to this issue and underlined the pessimistic image they have of the actions taken by companies in this regard. It is no longer a question of imagining 'greenwashing'[2] initiatives of questionable durability, but rather of rethinking our business models from A to Z. The objective of this chapter is to present five paths developed by Ipsos as part of its Trend Observer research program, plus one specific case study. These elements do not address the difficult question of rebuilding our economic models but rather show that concrete actions are possible and have already been taken by large international groups under the condition that they completely rethink their visions about the future (Prahalad, 2006; Friedman, 2007; Elkington and Hartigan, 2008). They shed some light on possible opportunities for development in the realm of sustainable innovation.

Consequently, this chapter will not develop possible directions for rethinking business models. Writing on that subject, by numerous researchers and consultants (Shrivastava, 1995; Martinet and Reynaud,

2004; Begg and Hart, 2005; Fiksel, 2009), is already quite extensive, and their various works have been followed by numerous debates. Our sole ambition here is to make a first attempt to classify the initiatives that have been taken around the world in terms of sustainable development. We will thus present a variety of trends that have been classified according to five distinct but non-contradictory conceptions of sustainable development. This classification is obviously debatable, and its only aim is to serve as a help in interpreting events. This vision is by nature fragmentary and partial, and is destined to evolve over time. But it can help business managers who are often short on time to position their brands and products in a fast-changing environment. The only postulate shared by all of these illustrations and the case study is that rather than focusing marketing questions on customers, they concentrate on individuals whose idiosyncratic behaviors have become the norm (Prahalad and Krishnan, 2008). In our view, this evolution of mass consumption toward the satisfaction of circumstantial needs and desires, where the brand plays a particular role in the customer relationship, is at the core of marketing's current interest in sustainable innovation. This innovation impacts both the conception of new products and the elaboration of new customer relationship channels entailing specific pathways and mixes. This dimension will be strongly highlighted by the following illustrations.

Before presenting the five paths developed by Ipsos and the specific case study of concrete marketing innovations in sustainable development, we want to focus on environmental new product development theory (NPD) in the academic literature, which is in accord with the different trends and cases we are going to present. We refer here to parts from the excellent D. Pujari, G. Wright and K. Peattie (2003) article about the building up of new product development theory towards environmental new product theory. But as seductive as this theory can be, we will see that it is not enough to answer marketers' questions about how to integrate sustainable development into company objectives.

Consumers' tastes are more and more fluctuating and it is becoming crucial for companies to understand how to shorten product life cycles (Calantone et al., 1995). This has made effective new product development theory (Pujari et al., 2003) to perform on the market for a few years. The essence of NPD lies in creating products whose core attributes (which deliver the basic benefits sought by customers) and auxiliary attributes (which help to differentiate between products) meet the needs of customers and other internal and external stakeholders. Selecting a good set of product attributes is achieved by a tightening process called 'Design-for-X' (Gatenby and Foo, 1990). Since 1990, a key 'X factor' has been 'design-for-environment', defined as 'a practice by which environmental considerations are integrated into product and process engineering design procedures' (Keoleian et al., 1995). This approach became essential for ten years from 1995 to 2005, with a peak in 1997 (Fuller, 1999). Responding appropriately to concern about the natural environment, the 'X factor' has changed many aspects of the way businesses operate and has become an integral part of purchasing, marketing, and corporate strategy (Wong et al., 1996; Shrivastava and Hart, 1994; Menon and Menon, 1997). The literature points to external benefits that arise from environmental improvement, including: increased sales (Fierman, 1991); improved customer feedback (Frankel, 1992); closeness to customers (Dean et al., 1995); enhanced competitiveness (Miles and Munilla, 1993; Porter and van der Linde, 1995); and improved corporate image (Engleberg, 1992; Kolk, 2000).

A specific NPD – the environmental NPD (ENPD) – is precise in its characteristics. ENPD is defined as product development into which environmental issues are explicitly integrated in order to create one of the least environmentally harmful products a firm has recently produced. This also includes the redesign of existing products to reduce their environmental impact in terms of materials, manufacture, use or disposal. Product performance is usually discussed in either technical or financial terms, often with a strong relationship between the two. Environmental concern has created the concept of eco-performance, which encompasses the socio-environmental impacts that a product has beyond the company and its marketplace (Peattie, 1995). In previous decades, if eco-performance was discussed at all, it was usually in terms of the trade-offs involved with technological or economic performance. Today, these three dimensions are increasingly viewed as interrelated. This new conception of ENPD has consequences for managers. They have to keep in mind 'A broader consideration of customer satisfaction'. Environmental concerns are leading to new customer requirements beyond conventional functionality, quality and cost, relating to how products are made, how long they last and how they can be disposed of (Peattie, 1999):

1. ENPD requires questions to be asked about the physical consequences of production and consumption. Addressing questions about where the raw materials going into products come from, and what happens to products post-use, reflect a physical 'cradle-to-grave' product life cycle perspective (Sharman et al., 1997).
2. A distinguishing feature of much ENPD activity is the attention given to the fate of products post-use, and particular design for the 'Five Rs' of repair, reconditioning, reuse, recycling and remanufacture (Wheeler, 1992).

3. While much early ENPD work employed a design-for-environment approach, which emphasized the reduction of the post-use environmental burden, more recently there has been an increased emphasis on the 'embodied' environmental burdens of the materials used (Simon et al., 2000). Suppliers have an important role in determining all aspects of product quality including eco-performance. ENPD requires a detailed understanding of the socio-environmental impacts of the whole supply chain, down to the simplest ingredient, which may previously have been perceived as standardized and unlikely to pose quality problems. Concern for the environmental impacts of suppliers can be seen in the introduction of the ISO 14 000 series of environmental management systems (EMS) quality standards to complement the ISO 9000 series. It can also be seen in the requirement of many businesses that their suppliers undergo environmental audits (Sinding, 2000).

But as we are going to note subsequently, things are getting more and more complicated and to gain a better understanding it is very important to identify trends to appreciate what 'sustainable development' really is for a marketer.

FIVE DIFFERENT CONCEPTIONS OF SUSTAINABLE DEVELOPMENT

Since 1997, the trends observer agency Ipsos has produced an annual analysis of major lifestyle and consumption trends around the world. This program goes by the name of Trend Observer. It focuses on the spread of innovative behavior, services and products and draws on international research. Fieldwork is conducted each year in France, the United Kingdom (UK), Sweden, Canada the US and Japan. The methodology combines several approaches: interviews with experts, interviews with a dozen or so trend-setters in each country,[3] and local and international trend-watching. The objective is to identify all types of new behavior, whether they stem from individuals themselves, from economic actors both private and public, or from interactions between new products, services and individual attitudes.

The year 2008 was a special year, one that Ipsos analysts consider to be 'year 1 of sustainable consumption'.[4] Even though the trend toward consuming products and services that are 'sustainable' and environmentally friendly existed years prior to the creation of Trend Observer in 1997, our research indicates that there was a genuine turning point in 2007. For the first time, there was strong convergence around this theme on a global level: sharpened awareness on the opinion level, consensus in the scientific community concerning global warming, apparent political resolve (the Grenelle Agreements, the Nobel Peace Prize awarded to Al Gore and the IPCC), and the multiplication of 'sustainable' initiatives in companies. We found evidence of this idea of entering a new cycle of consumption in all the countries studied, which reinforced the following Lester Brown conviction: 'In the area of food as well as in energy, we have left the age of abundance and entered the age of scarcity' (Brown, 2011). This is the reason Ipsos decided to devote a significant part of its 2008 Trend Observer report to sustainable development.

Our analysis of all the changes in consumption and the offer of products and services relating to sustainable development enabled us to identify five major trends. Each of these trends can be subdivided into several subtrends.

Trend 1: Back to Sources

This first trend expresses one of today's strong aspirations: getting back to the source. The idea, a reflection of Rousseau's ideas, is both simple and appealing: we will be able to advance down the road of sustainable development by returning to raw materials, to basic elements. Films such as *Into the Wild* directed by Sean Pean, and *The New World* written and directed by Terrence Malick, do a good job of illustrating this trend. In both films, and in very different contexts, getting back to nature – or a natural life – is presented as being the most constructive path to individual truth. This trend includes the following subtrends.

Use of raw materials

For those who are in favor of this trend, whatever is natural holds the greatest value. They place great value on raw materials drawn directly from the earth and the sea. In the broadest sense, this includes vegetable, animal and mineral. This is the reason the cosmetics business is one of the most active in this area. Several brands have launched products in the mineral and gemstone area in the past few years. The use of nacre by Skincare, diamond by Gemology and obsidian by Giorgio Armani's Crema Nera are just a few illustrations of this subtrend. Whereas these materials used to be excluded from cosmetic products because of the association one can have with 'hardness' and 'abrasiveness', the back to the sources trend now offers a favorable environment for innovation in the field.

We can cite a multitude of innovations stemming from this trend. For example, Ahava in London offers beauty treatments based on Dead Sea salts. In the furniture sector, Roche & Bobois, well known for its comfortable and sophisticated designs, presented its 'Legend' collection in 2007 based on the asymmetry and imperfection of tree branches. The brand Tasmanian Rain sells bottled rainwater in Australia and the US, with its ads presenting rainwater as being particularly 'pure' because it comes 'from the sky'. We can also include here the popularity in South Korea of 'Dr Fish' cafes, which have taken the secular practice of skin peeling by fish in special foot baths and updated it through a chain of fashionable cafes. This trend almost systematically places value on ancestral natural sources.

Importance of the local

The back to sources trend is not limited to materials. It also places value on locally produced products because transportation has become associated with a form of progress that is synonymous with pollution and contributing to global warming. It is assumed that emphasizing increased value placed on local production, out of a desire to respect natural growing cycles and to limit the pollution linked to the transportation of products, can only be beneficial. The US retail chain Whole Food Market, with its strong emphasis on maximizing its offer of local and seasonal food, is a good illustration of this subtrend.

Expansion of the natural concept

Last but not least, the natural, organic boom also expresses a need to get back to the source by using products that are 'pure', free from all types of pollution, and that respect seasonal cycles. A striking phenomenon in the past few years has been the extension of the organic beyond the food sector. From organic cotton children's clothes sold by Gap, H&M and Zara to toothpaste made by Tom's Maine and Colgate, no sector is left unaffected by the push to organic. Still relatively limited, this push corresponds to a desire for authenticity and traceability that considers the respect for and control over sources (materials, ingredients, etc.) to be a measure of safety and a source of well-being.

Trend 2: Voluntary Reduction

The second major trend we found aims to reduce the harmful and useless consequences of our consumption habits. It does not necessarily involve innovating, but rather reducing, combating the excesses of our mass consumption society. This panacea uses quantity as its measurement standard. According to this way of thinking, less can only be better. This trend includes the following subtrends.

De-pollution
The priority here is to reduce existing pollution. Two firms propose two ways to implement this subtrend: (1) modular systems of honeycomb-like architectural networks that reduce urban pollution are developed by Elegant Embellishments; and (2) depolluting vegetal walls that elevate plants from a decorative status to serving a genuine functional role in daily life by Wallflore in France. This evolution is highlighted in the work of the French designer Mathieu Lehanneur and his Bel-Air system, which also uses plants to clean the air and is designed to be installed indoors.

Voluntary simplicity
Beyond the fight to reduce pollution, this trend also implies so-called 'reverse growth'. This principle draws on 'voluntary simplicity': eliminate the superfluous and concentrate on the essential. The novelty resides in the fact that this principle, defended for decades by numerous ecology activists, is now being claimed and adopted by people working in fashion and industry. Thus, the clothes designer Vivienne Westwood declared during Fashion Week in Sao Paulo in 2008: 'I'd like to see people buy less, focus on quality items. Sometimes I want to ask my manager to let me create less each season so that I can concentrate more on each piece.' Numerous young designers follow this credo. Designers talk about the concepts of 'ethical fashion' and 'sustainable fashion'. In the automotive field, Citroën's C-Cactus concept car is based on these essential values. It reduces the car as much as possible to its basic functions and only retains what is useful: there is no dashboard, no body paint or varnish. But this model is not 'sad'-looking at all. On the contrary, its design is modern and attractive. We can reduce the materials used and rationalize functions while at the same time embellishing appearance: a way to reconcile the principles of pleasure and responsibility.

Recovery, recycling
In this 'reduction' way of thinking, waste becomes raw material. The ancient practice of recycling is part of this reduction effort. An increasing number of large businesses have adopted this theme and made it the spearhead of their sustainable strategy. Vodafone, like nearly all of today's mobile phone operators, recovers and recycles old mobile phones. Creativity now calls for reinvention, just like the Nike Considered shoes from Nike whose outer soles are made from rubber residues. At the end of 2008, Quicksilver launched its Eco-Design line made with recyclable Eco Circle fibers. Items in this line will be recovered by stores and recycled as new clothes.

Revival

Reduction can also be expressed through the revival of older 'natural' techniques. This has led to the revival of natural methods of energy generation. The hand generator flashlight from TTP works without batteries and can also be used to charge mobile phones. The 'crank' or 'pull-string' computer that was once intended to spread computer use in Africa works with a hand crank generator. In 2007 Sony presented a prototype of digital still camera in the form of a wheel attached to a handle, without batteries. A picture can be taken after rolling the wheel by hand for 15 seconds to generate sufficient electricity. In each case, the innovation consists of using natural sources of energy, which make it possible to become independent of costly or defective energy networks.

Trend 3: Ethical Concerns

Another way to consider sustainable development is through ethical concerns. Likely less productive but calling for a great deal more responsibility, this conception of sustainable development considers individuals and enterprises as fully responsible for building the future. This trend includes the following subtrends.

Transparency

The first illustration of the ethical approach is the willingness to be transparent with consumers. The number of initiatives in this area has increased enormously in the past few years. One of the best known is undoubtedly SNCF's EcoComparateur. It provides the volume of CO_2 emitted during a given trip with different transportation options. New product labels being tested by the grocery retailer Casino in collaboration with Bio Intelligence Service inform consumers about the complete life cycle environmental impact of the products they buy. Implementation of this type of labeling was, in fact, one of the measures adopted as part of the Grenelle Agreements. Another example is the Power Aware Cord, an electric cord that gives users a visual indicator of how much electricity they use. The cord glows blue when in use. The higher the electricity demand of the appliance that is connected to it, the brighter the cord glows.

Financial ethics

The financial sector has also launched consumer banking products to promote environmentally friendly investments. This concept is called ethical financing. The green bank cards launched by the MBNA Bank in Canada make it possible to earn points when making 'green' purchases with the cards. BNP Paribas launched a personal loan called Energibio dedicated to energy-saving home improvements. The Caisse d'Epargne also launched a green A-type passbook account, and this initiative will be extended to all banks in France through the implementation of Sustainable Development passbook accounts.

Commitment

Companies are also willing to impose restrictive rules on themselves (and also on their suppliers and customers) in order to limit their environmental impact. As such, the Monoprix retail chain has committed itself to transporting its merchandise in more ecologically friendly ways by, for example, using barges for long trips, local delivery vans powered with natural gas and a delivery service that only delivers once the vans are full. Certain brands have gone as far as adopting strong activist messages, perhaps in reaction to advocacy writing such as Michael Norton (2007). Yves Rocher, for example, launched its 'Liberté, égalité, beauté' campaign at the end of 2007.

The subtrends we have looked at to this point all express, in one way or another, a concern about responsibility. Seen from the outside, this concern can appear to be somewhat serious, restrictive, at odds with the pleasure principle our consumer society is based on. But there are other 'pleasurable' ways to act in support of sustainable development. These are the themes we have grouped together under the trend 'guilt absolution'.

Trend 4: Guilt Absolution

'Guilt absolution' includes ways to reconcile the pleasure principle and the responsibility principle. In this case, ecology is not a commitment or a necessary duty: it becomes 'valorizing'. The environmental imaginary becomes fashionable. In an example in Ginza in Japan, the Anya Hindmarch boutique provoked a stampede when it put its 'I am not a plastic bag' bags on sale in 2007. The product was environmentally friendly and became a fashion object that was a particular favorite 'it-bag' for fashionistas. Another example is the French company Reversible which recycles advertising tarps, turning them into fashion accessories. They have been successful and their products, most of them hand and shoulder bags, have become 'must-have' items among the trendy population. It is not important whether or not these successes last: they combine pleasure and responsibility and enable fashion-hungry consumers to 'absolve guilt'. What is essential is to bring together the logic of responsibility and the logic of fashion. A final example: in New York, Barney's 2008 Christmas window was decorated with recycled soft drink cans.

People can also 'reimburse' their debt to nature. The spread of carbon offset schemes is developing through organizations such as Climat Mundi and the Carbon Neutral Company. This approach is again another way for consumers to 'absolve guilt'. Trend Observer showed that the carbon offset schemes helped certain scooter owners feel less guilty. They feel better about the idea of continuing to do something they enjoy (driving a scooter) while also paying their debt to nature.

Trend 5: 'Holistic' Green

The 'holistic' green is different from all the others. We call it the 'holistic green' trend. It does not follow a logic of continuity starting with what exists. It does not try to navigate a transition between our current way of life and the behavior of the eco-citizens of the future. It adopts a position of total or even abrupt change. The objective is to take a global approach that is environmentally friendly across all dimensions. Today, it is firmly established in architecture and urbanism, and it aspires to spread itself to other sectors. A number of experimental projects that rethink new ways of living together have been or are soon to be completed around the globe.

New York inaugurated its first 'green building' in spring 2008, which brings together a shopping complex and a 'green' hotel: the Greenhouse 26 in Chelsea. Europe is not being left behind and has a 'green' building of its own in progress since 2009, the Tour Phare in Paris's La Defense area. Presented as a model of its type, with wind generators on the roof and a double-skin exterior making it possible to regulate heating and cooling, this new-generation tower proposes an innovative concept of vertical living.

The company LivingHomes proposes pre-fabricated 'green' homes that it has built for customers in a variety of places. Located south of Stockholm, the Hammarby Sjöstad district claims to be the largest green neighbourhood in the world. All the housing is equipped with triple-pane windows, low water consumption toilets and is built with insulating materials. Electrical appliances are furnished and are chosen for their low energy consumption qualities. The neighbourhood is also a model of waste recovery: wastewater is treated locally, with the sludge used to produce biogas to heat the apartments and power city buses.

Genuine eco-cities are also starting to appear, such as Dongtan in China which opened in 2010 and, more notably, Masdar City close to Abu Dhabi. This environmentally friendly city, without cars or skyscrapers and with zero CO_2 emissions, is scheduled to welcome 50 000 residents and 1500 businesses starting in 2016. It will cover 6 km² and is expected to cost $22 billion. It will also be the future home of IRENA, the International Renewable Energy Agency.

All of these examples present a 'holistic' dimension: a quest for sustainable development that omits nothing. All steps of the process and all dimensions of the 'product' should be subject to a 'green' treatment.

In conclusion, there are multiple marketing visions of sustainable development that are gradually taking form in customer promises. As we said earlier, these five conceptions of sustainable development are not intended to provide a thorough vision of the situation or be a turnkey manual of sustainable development for managers. Nevertheless, these are real and effective illustrations (they are not projects, but actual products and services being sold to customers) that provide an indication of the amplitude of the movement and offer marketing professionals a help for ongoing reflection.

Now that we have identified trends and subtrends for marketing innovations in sustainable development, we think that it is interesting to study a specific case. We are going to do that by presenting the innovating Web 2.0 website 'Ma maison Bleu Ciel d'EDF'.

SUSTAINABLE DEVELOPMENT AS AN INNOVATIVE SUPPORT TO ENERGIZE CUSTOMER RELATIONSHIP MANAGEMENT

We previously saw five different marketing trends in sustainable development, all of them based on new product concepts. However, marketing does not only consist in the conception of new offers, but also in customer relationship management (CRM). CRM presents several opportunities to deeply modify merchant and social relationships.

The utility EDF Group, with its more than 20 million residential customers in France, tried to introduce the concept of sustainable development in its CRM by launching the initiative Web 2.0 website 'Ma maison Bleu Ciel d'EDF'.

Ma Maison Bleu Ciel d'EDF (My EDF 'Bleu Ciel' House[5]) is a virtual house that can be used on the Web at no charge but that has private access, enabling EDF to experiment with a new form of customer relations based on Web 2.0 technologies. This is the first time that EDF has given its residential customers the opportunity to express themselves on a blog, a forum and a 3D interface on the theme of energy eco-efficiency.

Objectives

This website aims at experimenting with a new form of customer relationship that is both customized and community-based in the context of a simulated house. The Internet users can create their house within a village where they can be in contact with the other inhabitants. This is a village in which each 3D house has all the main rooms: dining room, kitchen, living room, bathroom, garage. The users can personalize the rooms as much as they want: more than 300 pieces of furniture and 100 appliances are available.

The home equipment supports an array of services with their own specific uses that enables the user to be connected to reality, combining both pragmatism and simplicity: a weather report for the EDF contract location on the windows, seeing a video or watching television programs on TV, reading the headlines of the national daily newspapers in the newspaper rack, and so on.

The Bleu Ciel d'EDF brand promise, 'more savings, more ecology, more well-being', is furthermore evidenced by three gauges that react immediately any time equipment is added or deleted. This consequently has an impact on the power consumption of the room, the CO_2 emissions (when applicable), and the notion of comfort.

The users are invited to give their opinion on the calculation of these gauges, for any part of the house or the whole house itself. They can share their opinions with other users, visit other people's houses, invite their friends, create challenges, and so on. In addition, they can simulate the power consumption of their electric appliances, follow personalized advice in terms of eco-efficiency or become involved in support of sustainable development in their day-to-day lives.

Results Delivered or Targeted

The experimentation was launched in November 2008. In 2011, more 37000 houses had been created and 1500 comments posted. But some comments have been read more than 1300 times. This experiment not only enhances the relationship between the end users and the company but it also gives the brand a human touch by receiving the opinions of customers.

At the same time, it gives the marketing department the opportunity to be able to better understand the customers' expectations in the area of sustainable development within the setting of their virtual homes. This way, they can tap into new marketing opportunities by proposing ideas for new offerings to the most frequent users – since the houses are virtual but the behaviours and needs reflected in them are real – to obtain their opinions and create these offerings with them. If consumption per use were to develop, the house could also be an interface for the high-quality managing and educational showcasing of the measures carried out.

Ma Maison Bleu Ciel d'EDF is merely a tool that can prove to be a medium rounding out the traditional channels for gathering community input and establishing a high-quality, automated multimedia customer relationship. The tool is highly flexible and can also be adapted to the use that the company wants it to have: a day-to-day customer relationship tool, education on eco-efficiency, an interface for steering the consumption of appliances in the home, a communication vector for debates, and so on.

This innovation also offers the advantage of not having a large number of elements to be translated in text form for an international company with numerous subsidiaries. The simulators all work in the same way; only the equivalence tables need to be adapted and the customers' needs are the same everywhere: cut down on one's energy consumption while benefiting from a higher level of comfort without sacrificing the planet's resources. The action is therefore easily transferable to the subsidiaries of the group that express their interest, while retaining the centralized research and development (R&D) practices in the home base. The marginal cost for disseminating the product is close to zero once the translation and the choice of content are completed.

These technologies draw upon a virtual customer relationship to obtain standardized or multimedia-based information. Ma Maison Bleu Ciel d'EDF is therefore not only a source of savings, but also creates value since it focuses on a high-value-added customer relationship. The house provides answers to the initial before-sales or day-to-day management questions and refers customers to the phone or shops if they require more in-depth information. By developing this new form of customer relations, the Bleu Ciel d'EDF brand is exposed to the customer in a concrete and innovative way, which enables it to save money on advertising.

To conclude, the question of sustainable development brings more sources of innovations than it causes constraints for companies, at least from a marketing point of view. However, this is just the beginning of a worldwide phenomenon which is going to change our standard of living. This is not the first time in mankind's history that social shifts have been led by trade and new needs. But it is the first time at a worldwide scale that marketing will take an initial role in promoting innovative standards of living, due to its segmentation and targeting approaches. Regulations and products are just consequences of brands' decision policies: only brands are at the center of this day-to-day concern. We hope that brands will know how to grab these opportunities and that marketing directors will be able to convince stakeholders that there is no discrepancy between social shifts and benefits.

NOTES

1. Forty-six percent of the French feel that their companies do little or nothing for the environment. Nevertheless, it is an issue that has captivated their attention: only 8 percent of French interviewees say they are not interested in it.
2. Companies are still not completely credible as 'green': 8 percent of interviewees feel companies engage in greenwashing. Criticism increases with age: this answer was given by 4 percent of interviewees aged 18–24, 10 percent of those aged 25–34, 13 percent of those aged 35–54 and 33 percent of those aged 55 and over (LinkedIn, 2009).
3. Definition of a trendsetter according to Ipsos: an individual aged 25 to 35, professionally active, with an 'open' lifestyle, alert to new ideas. The psycho-sociological profile, which is extremely difficult to establish, ensures the collection of quality information: new, based on direct experience, tied to innovation.
4. '2008 is emerging as "year 1 of sustainable development", a year in which a number of strategies will be implemented on the collective and the individual levels to help the planet endure. Companies, political entities and civil society are more deeply involved in "green" innovation, but individuals aren't standing on the sidelines. They are increasingly adopting the role of eco-actors of their own lives, even becoming active producers of their own resources', from Trend Observer 2008, Ipsos Marketing.
5. For more details see: www.mamaisonbleucieledf.fr.

REFERENCES

Begg, C. and S. Hart (2005), *Capitalism At The Crossroads: The Unlimited Business Opportunities In Solving The World's Most Difficult Problems*, Upper Saddle River, NJ: Wharton School Publishing.

Brown, L. (2011), *Plan B 4.0: Mobilizing to Save Civilization*, London: W.W. Norton & Co.

Calantone, R.J., S.K. Vickery and C. Droge (1995), 'Business performance and strategic new product development activities: an empirical investigation', *Journal of Production and Innovation Management*, **12**(3): 214–26.

Dean, T.J., D.M. Fowler and A. Miller (1995), 'Organizational adaptations for ecological sustainability: a resource-based examination of the competitive advantage hypothesis', Working Paper, Department of Management, University of Tennessee, Knoxville, TN.

Elkington, J. and P. Hartigan (2008), *The Power of Unreasonable People: How Social Entrepreneurs Create Markets That Change the World*, Boston, MA: Harvard Business School Press.

Engleberg, D. (1992), 'Is this the best community relations program in the country?' *Environment Manager*, **3**: 6.

Fierman, J. (1991), 'The big muddle in green marketing', *Fortune*, **123**: 91–101.

Fiksel, J. (2009), *Design for Environment: A Guide to Sustainable Product Development: Eco-efficient Product Development*, 2nd edn, Boston, MA: McGraw-Hill Professional.

Frankel, C. (1992), 'Blueprint for green marketing', *American Demographic Journal*, **14** (4): 34–8.

Friedman, T.L. (2007), *The World is Flat: The Globalized World in the Twenty-first Century*, New York: Penguin Books.

Fuller, D. (1999), *Sustainable Marketing: Managerial–Ecological Issues*, Thousand Oaks, CA: Sage.

Gatenby, D.A. and G. Foo (1990), 'Design for X (DFX): key to competitive, profitable products', *AT&T Technical Journal*, **69**(3): 2–13.

Keoleian, G.A., J.E. Kock and D. Menerey (1995), 'Life cycle design framework and demonstration projects', US Environmental Protection Agency Report, Washington, DC.

Kolk, A. (2000), 'Green reporting', *Harvard Business Review*, **78**(1): 15–16.

LinkedIn (2009), 'Press release for World Environment Day', 5 June.

Martinet, A.C. and E. Reynaud (2004), *Stratégie d'Entreprise et Ecologie*, Paris: Economica.

Menon, A. and A. Menon (1997), 'Enviropreneurial marketing strategy: the emergence of corporate environmentalism as market strategy', *Journal of Marketing*, **61**(1): 51–67.

Miles, M.P. and L.S. Munilla (1993), 'The eco-orientation: an emerging business philosophy', *Marketing Theory and Practice*. **1**(2): 43–51.

Norton, M. (2007), *The Everyday Activist: Everything You Need to Know to Get Off Your Backside and Start to Make a Difference*, London: Boxtree.

Peattie, K. (1995), *Environmental Marketing Management: Meeting the Green Challenge*, London: Pitman.

Peattie, K. (1999), 'Trappings versus substance in the greening of marketing planning', *Strategic and Marketing Journal*, **7**(2): 131–48.

Porter, M.E. and C. van der Linde (1995), 'Green and competitive: ending the stalemate', *Harvard Business Review*, **73**(5): 120–133.

Prahalad, C.K. (2006), *The Fortune at the Bottom of the Pyramid*, Upper Saddle River, NJ: Wharton School Publishing.

Prahalad, C.K. and M.S. Krishnan (2008), *The New Age of Innovation: Driving Cocreated Value Through Global Networks*, Boston, MA: McGraw-Hill Professional.

Pujari, D., G. Wright and K. Peattie (2003), 'Green and competitive influences on environmental new product development performance', *Journal of Business Research*, **56**(8): 657–71.

Sharman, M., R.T. Ellington and M. Meo (1997), 'The next step in becoming "green": lifecycle orientated environmental management', *Business Horizon*, **40**(3): 13–22.

Shrivastava, P. (1995), *Greening Business: Profiting the Corporation and the Environment*, Cincinnati, OH: Thomson Executive Press.

Shrivastava, P. and S. Hart (1994), 'Greening organizations – 2000', *International Journal of Public Administration*, **17**(34): 607–35.

Simon, M., S. Poole, A. Sweatman, S. Evans, T. Bhamra and T. McAloone (2000), 'Environmental priorities in strategic product development', *Journal of Business Strategy*, **9**(6): 367–77.

Sinding, K. (2000), 'Environmental management beyond the boundaries of the firm: definitions and constraints', *Journal of Business Strategy*, **9**(2): 79–91.

Wheeler, W.A. (1992), 'The revival of reverse manufacturing', *Journal of Business Strategy*, **13**(4): 8–13.

Wong, V., W. Turner and P. Stoneman (1996), 'Marketing strategies and market prospects for environmentally-friendly consumer products', *British Journal of Management*, **7**(3): 263–81.

Index

3M Corporation, Pollution Prevention Pays Program (3P) 12

Abrassart, C. 186
Abu-Lebdeh, Ghassan 204–18
Ackermann, C. 146
Acquaye, A. 206
Africa 152, 155, 156–7, 158, 160
Afuah, A. 124, 128
Aggeri, F. 186
Agrawal, A. 92
Akinboade, O. 159
Al-Mansi, A. 252
Alan, C. 191
Alan, K. 191
Aldrich, H. 152
Alter, S. 139, 140, 148
Alvord, S. 173
Ambec, S. 65–6
Amul milk cooperative 131–2
Anderson, A. 141
Andres, P. 93
appropriate technology movement 118–35
   Buddhist economics 123–4
   competitive restraints 129–30
   dehumanising nature of industrialisation 121–2, 124
   early industrialisation 118–19
   economic growth, importance of 128–33
   India and early iron technology, effects of 122
   India, HMT rice variety, development of 125–6
   Indian cultural idiosyncrasies and business decision making 131–3
   Indian Freedom Movement and factory system 119–20
   innovation dropout problem 129
   innovation and invention,
      commercialisation and marketing issues 126–8
   innovation and invention increase 121, 124, 125–8
   intermediate technologies 122
   micro systems, importance of 127–8
   National Innovation Foundation, India 126–7
   online marketing and open source initiative, India 127
   outsourcing, benefits for India 130–31
   production efficiency measurement and profit motive 120–21
   religion and philosophy, effects of 123–6
   'scanning the environment' method 130
   social impact 125–6
   sustainability factors 124–5, 127
   Taylor System 120
Arning, J. 253
Atkinson, G. 55, 144
Augenbroe, G. 208–9
Austin, J. 148
Ayuso, S. 70

Babu, S. 149
Balasubramanian, N. 93
Bar-on, A. 159, 160
Bardy, Roland 139–67
Barney, J. 186
Barney Pityana, N. 152
Barreyre, P. 188
Basant, R. 77
Batsch, L. 229
Bavadam, L. 125
Baysinger, B. 92
Beamon, B. 221
Becker, D. 223
Begg, C. 263
Beheiry, Salwa 204–18

Bekker, G. 213
Bellini, B. 190, 194
Bennett, S. 141
Berger-Douce, Sandrine 186–203
Berle, G. 141
Berry, M. 13
Bhagat, S. 92
Bhāle, Sanjay 168–85
Bhāle, Sudeep 168–85
Binns, T. 157
Bioly, Sascha 239–61
Black, B. 92
BMW Group, eco-innovation 32
Boissin, J. 187, 191
Bornstein, D. 160
Borzaga, C. 142, 148–9
Bowen, P. 205
Bowersox, D. 220
Boyko, C. 210
Brake, K. 253
Braungart, M. 145
Brazil 153, 177
Bridwell, L. 207
Brinckerhoff, P. 177, 178
Brown, K. 124, 155
Brown, L. 266
Browne, F. 252
Brundtland, G. (Brundtland Report) 3,
　　52, 58, 220
Buckley, P. 147
Butler, H. 92
Butynski, T. 157
Bynum, P. 209–10

Calantone, R. 263
Callard, A. 157
Camison-Zomosa, C. 186
Campbell, K. 158
Carrier, C. 188–9
Carrillo-Hermosilla, J. 48
CEMEX 148
Champy, J. 220
Charter, M. 26
Christensen, C. 128–30
Christopher, M. 220, 222
Clark, T. 26
Clements, M. 208
climate change *see* global climate
　　change, sustainable innovation
　　responses to

Clinton, S. 220
Closs, D. 220
Coase, R. 66
Cohen, B. 141
Cole, A. 171
competitiveness
　　regulatory standards and social
　　　　challenge of sustainable
　　　　development 64–7
　　restraints, appropriate technology
　　　　movement 129–30
construction technology in the building
　　and transportation sectors,
　　benchmarking, building sector
　　205–10
　　building information modeling
　　　　(BIM) 209–10
　　and energy consumption 206
　　environmental policies, need for
　　　　208–9
　　flexible building design 206–7
　　fundamental safety risk levels 209
　　future visualization tools 210
　　green building assessment and
　　　　measurement tools 207
　　Leadership in Energy and
　　　　Environmental Design (LEED)
　　　　credits 209
　　life-cycle assessment (LCA) and
　　　　environmental impact 208
　　off-site fabrication and alternative
　　　　contracting strategies 206–7
　　SILENT sustainability assessment
　　　　model 209
　　solar technology 207
　　sustainability in developing countries
　　　　208
　　sustainable construction, increased
　　　　popularity of 207–8
　　Sustainable Construction
　　　　Technology Index case study
　　　　214–15
　　team communication, importance
　　　　of 209
　　Terra Block Fabricator 207
　　waste management programs 208
construction technology in the building
　　and transportation sectors,
　　benchmarking, transportation
　　sector 210–14

air-pollution-absorbing concrete and
    bricks 213
global warming effects 212–13
high speed rail (HSR) and energy
    efficiency 212
low-energy-consuming materials 212
natural habitat impact reduction 213
noise pollution 213
project operation 210–11
recycled materials, use of 211–12
sustainability policies, need for 214
unsustainable projects, identifying
    impact of 211
consumers *see* customer relationships
Convention on International Trade in
    Endangered Species (CITES) 8
Conway, S. 129
Cooper, M. 227
Corner, P. 176
corporate governance norms and
    Indian corporate enterprises 74–9
Accounting Standards 80
auditor–company relationship 78,
    81–2, 85–7, 90
banks and financial institutions,
    supervisory role of boards 79
Birla Committee (Securities and
    Exchange Board of India
    (SEBI)) 78, 91, 94
board composition and
    independence 83–5, 92, 93–4
Companies Act 75, 77, 79, 80, 84–5,
    86
Confederation of Indian Industries
    (CII) initiative 77–8
Confederation of Indian Industry
    Code on Corporate Governance
    93–4
Constitutional Law and wealth
    distribution 79
corporate governance overview 74
disclosure transparency 78–9
evolution of 75–7
future research 92–5
Ganguly Committee (Reserve Bank
    of India (RBI)) 79
Income Tax Act 80
Indian Partnership Act 75
information and reporting pattern,
    inadequacy of 90

initiatives in corporate governance
    77–9
insider trading and mergers and
    acquisitions 91
Institute of Chartered Accountants
    of India (ICAI) 80, 81, 87
institutional context and firm
    performance 93
institutional investors, monitoring
    role 87–90
Irani Committee 79, 94
liability for independent directors 94
liberalization effects 77
Malegam Committee (Securities
    and Exchange Board of India
    (SEBI)) 78
mergers and acquisitions 75, 90, 91
Ministry of Corporate Affairs
    (MCA) 79, 80, 81, 87
Naresh Chandra Committee
    (Department of Company
    Affairs) 78, 94
N.R. Narayana Murthy Committee
    94
post-1990 period 77
post-independence period 76–7
poverty 91–2
pre-independence period 75–6
regulatory framework 79–80, 87,
    94–5
Restrictive Trade Practices Act
    (MRTP Act) 77
sanctions and enforcement for
    violations 87
Securities and Exchange Board of
    India (SEBI) 77, 79–80, 81, 87,
    88–9, 91
Securities and Exchange Board of
    India (SEBI), Takeover Code 90
shareholder activism, need for 87–8,
    94–5
shareholding of controlling interests,
    need for identification of 90
sustainable solution, need for 91–2
violations, implications of 87–91
see also marketing innovation;
    multinationals, codes of
    conduct and other multilateral
    control systems for
corporate governance norms and

Indian corporate enterprises,
Satyam Computer Services case
study
  auditors' role 85–7
  board composition and
    independence 83–5
  corporate governance norms,
    transgression of 83–5
  PricewaterhouseCoopers (PWC)
    audit 81, 82, 86
  unearthing of scam 82
Courrent, J. 191
Craig, P. 174
Cramer, W. 6
Cuervo-Cazurra, A. 109
customer relationships
  consumer choices, importance
    of, and entrepreneurship
    development at small scale
    182–3
  customer relationship management
    (CRM) 263–4, 272–4
  eco-innovation and green industry
    transformation in OECD
    countries 42–4
  transparency and marketing
    innovation, Ipsos analysis 269

Dalton, D. 93
Daly, H. 52, 55, 144
Damanpour, F. 188
Das, N. 77
Das Gupta, J. 155
Daval, H. 187, 191–2, 199
Davidsson, P. 175
Davis, J. 186
Davis, M. 253
De Soto, H. 146
Dean, T. 264
Dees, J. 173
Delfmann, W. 244
Denmark 14, 15, 44
developing countries
  construction technology in building
    sector 208
  eco-social business *see* eco-social
    business in developing countries
  environmental protection, challenge
    of 52–3
Dewberry, E. 220

Dewlaney, S. 209
Di Maggio, P. 190
Dickson, B. 141
Diver, S. 158
Do Prado Lima, G. 153
Dobson, A. 145
Dornier, P. 228
Dorward, A. 142
Dosi, G. 200
Drucker, P. 169, 170, 172, 175
Ducuing, O. 201
Dunay, R. 207
Dunning, J. 101, 109, 110
Dur, F. 209
Dyllick, T. 145

Easterly, W. 150
eco-innovation and green industry
    transformation in OECD
    countries 21–50
  air conditioning 36
  alternative business models, use of
    33–5, 39
  bicycle sharing system 32, 34
  business models, conceptualisation
    of 39–44
  closed-loop production system 28–9
  customer relationships 42–4
  demand-side policies 44–5
  disruptive innovation 30, 31–2, 38
  eco-industrial park initiatives,
    requirements of 29
  eco-innovation definition 26–7
  economic effects 23, 38
  electronics industry 32, 33–4, 36
  environmental innovations, intended
    and unintended 26
  environmental technologies,
    emphasis on 24–5
  future vision, need for 47
  GDP growth decoupling 23, 38
  general-purpose technologies 38–9
  good practices framework 32–7
  green growth, innovation for 24–8
  green growth as new policy
    crossroads 22–4
  greenhouse gas (GHG) emissions
    23–4, 33, 35, 38, 44–5
  incremental innovation 29–30, 31–2,
    38

industry practices, understanding 28–32
information and communication technologies (ICT) 33, 38–9, 45–6
innovation mechanism 27–8
iron and steel industry 32, 33, 35
levels of making differences, classification of 26
physical infrastructure support, availability of 46
policy considerations 44–6
product-service systems (PSSs) 39, 40, 42
radical forms of eco-innovation, importance of 29–30, 31–2, 38, 45
risk considerations 31
shared products 42–4
social and cultural changes, effects of 31
social and institutional structures 26
sustainable development targets 24–5, 28–9
systemic changes and business opportunities 37–9
systemic (transformative) innovation 30–32, 45
technological entry barriers 45
transport industry 32, 33, 35, 38, 44–6
value creation 39, 41, 43
eco-logistics improvement in France 219–38
business process re-engineering (BPR) 220
freight transportation example and sustainable development 221–3
logistics development, role of transport and territorial policies 227–8
logistics, integrated and sustainable 220–21
logistics networks, collaborative aspect and global geographical dimension 229, 234–5
logistics networks, multiplicity of 221–3
oil prices, effects on transportation choices 225

service level improvements 228
supply chain networks 219–23
sustainable development definition 220
transportation modes, cost specifications 232
eco-logistics improvement in France, inland waterway transport advantages of 233
'door-to-door' service and delivery deadlines 235–6
high-value-added goods transportation 226
integration, barriers limiting 233–4
integration, key to successful 234–5
integration in logistics networks 229–31
integration, and new practices and strategies 233–6
integration, speeds limitations 233–4
integration within supply chains, effects of 227–33
logistics networks, ecological benefits 230–31
logistics networks and lower transport costs 229–30
logistics risks and external costs, reduction of 231
new commercial offers 235–6
recent increase in, cyclical and structural factors 225–7
recent increase in 224–5
revival of 223–7
revival of, regulatory changes 223, 225
as transportation mode dedicated to flexible production system 231–3
wide-gauge canal projects 224–5, 227
eco-social business in developing countries 139–67
barriers to 142
bottom/base of the pyramid (BOP) strategies and public purpose capitalism 146–8, 150
business entrepreneurship 150
Climate and Biodiversity Convention (CBD) 153–4
community-based participation and expanded linkages 152–3

eco-tourism businesses and cross-
    border investment 157
empirical cases, model application
    155–60
enterprise development funds 147
entrepreneurs, role of 140, 141–3,
    150
factor-four strategy and recycling
    145–6, 151–2
foreign investment by business firms
    155–60
fully indigenous eco-social
    businesses 157–8, 159–60
global markets and co-venturing
    146–7
green and clean technology
    implementation 152
informal sector, inclusion of 146
information technology effects 142,
    155
linkage phenomenon 147–8, 152–3,
    155
literature review 141–6
living conditions and natural
    environment improvements
    147–8
local power structures, effects of
    143
low-carbon FDI 155–6
NGO involvement 156
nonprofit sector and market-based
    solutions 139
not-for-profit institutions and
    financial self-sufficiency 148–9
policy entrepreneurship 149
policy frameworks 153–5
poverty reduction and business
    entrepreneurship 150, 153,
    155–6
poverty reduction and foreign
    investment 147–8
program entrepreneurship 149–50
public purpose capitalism and
    investment spillovers 147
science-based businesses and foreign
    R&D 156–7
social enterprise definitions 139
social entrepreneur mission 149–50
sustainability and collective action
    145–8

sustainability issues and responsible
    investment 143–6
sustainability, sink-side and source-
    side problems 151–2
theoretical foundation 148–55
transnational companies (TNCs),
    involvement of 155–6
upcycling and downcycling 145
weak versus strong sustainability
    and natural capital replacement
    144–5
World Conservation Strategy (WCS)
    154
youth and youth leaders as social
    entrepreneurs 149–50
see also entrepreneurship
    development at small scale as
    key to sustainable economic
    development
ecological perspective
logistics management 246–8
social challenge of sustainable
    development through
    innovation 63–4
economic effects
eco-innovation and green industry
    transformation in OECD
    countries 23, 38
economic crisis and social challenge
    of sustainable development
    through innovation 54
economic growth, challenge of 53–6,
    60–61, 63
economic growth, importance
    of, appropriate technology
    movement 128–33
financial ethics, Ipsos analysis
    269–70
green economy *see* green economy
logistics management, integrating
    sustainability and technology
    innovation, Germany 241–6
sustainable development *see*
    entrepreneurship development
    at small scale as key to
    sustainable economic
    development
Ede, F. 191
Edwards, M. 142
Ehrenfeld, J. 3, 9

Ehrlich, A. 10, 63
Ehrlich, P. 10, 63
Eickmann, C. 241
EID Parry 148
Eisenberg, T. 93
Eisenkopf, A. 241
Ekins, P. 63
electronics industry 32, 33–4, 36
Elram, L. 227
Elsen, T. 158
Engleberg, D. 264
entrepreneur profile and sustainable
     innovation strategy 186–203
   entrepreneurial profile 191–2, 194,
      196–9
   environmental technologies and
      sustainable development 190–91
   French eco-activities and Grenelle de
      l'Environnement 189–90
   innovation and entrepreneurship
      187–9
   innovation and performance,
      relationship between 188, 195,
      196, 197
   managerial issues 200–201
   sustainable innovation strategies
      189–91, 193–4
   technological innovation,
      importance of 188–9
entrepreneur profile and sustainable
     innovation strategy, ECODAS
     case study, France 193–9
   entrepreneur's profile, influence of
      196–9
   environmental regulation, evolution
      of 199
   intellectual rights 197–8
   non-discriminatory employment
      195–6, 198–9
   sustainable development strategy
      194–6
   sustainable development and
      technological innovation 193–4
entrepreneurship development at
     small scale as key to sustainable
     economic development 168–85
   business entrepreneurship definition
      171–2
   consumer choices, importance of
      182–3

entrepreneurial attitude and
      innovation 170–71
   India, Lijjat Papad cooperative
      example (SMGULP) 178–80
   India, Lijjat Papad cooperative
      example (SMGULP), business
      model 180–81
   India, Lijjat Papad cooperative
      example (SMGULP), value
      system 181–2
   innovation and realization 169
   Internet access 177
   Millennium Development Goals
      (MDGs) 178
   opportunity recognition 172, 174–5
   poverty solutions 171, 175, 176–7
   private sector involvement and
      growing importance of SMEs
      182–3
   profit-making and social
      entrepreneurship 175, 176, 181
   value creation 172, 173, 174–8, 181–2
   see also eco-social business in
      developing countries
entrepreneurship development at
     small scale as key to sustainable
     economic development, social
     entrepreneurship 170–71
   definition 173–5
   key features 174–8
   socially entrepreneurial ventures
      (SEVs) 173–4
   and sustainability 175–6
   versus business entrepreneurship
      171–3
   vision-oriented and crisis-oriented
      factors 175
environmental concerns
   carbon offset schemes 270–71
   environmental balance and electric
      mobility 254–6
   environmental technologies,
      emphasis on 24–5
   externalities factor and social
      challenge of sustainable
      development through
      innovation 66
   policies, need for, and sustainable
      construction technology in
      building sector 208–9

protection in developing countries,
   challenge of 52–3
recycling 14, 145–6, 151–2, 211–12,
   264, 268
regulation, evolution of, ECODAS
   case study, France 199
see also 'eco' headings; global
   climate change, sustainable
   innovation responses to; green
   economy
EU
   City–Vitality–Sustainability (Civitas)
      initiative 214
   Eco-Innovation Action Plan 24
   Eco-Innovation Observatory (EIO)
      project 26
   European Waste Directive 199
   Lisbon Strategy for competitiveness
      and economic growth 24
   White Paper on European Transport
      Policy 227–8
Evans, D. 212

Faucheux, S. 55
Fayolle, A. 187
Fender, M. 228
Fierman, J. 264
Fiksel, J. 263
financial effects *see* economic effects
Fincham, R. 129
Fisher, M. 221
foreign investment *see* eco-social
   business in developing countries
Foster, R. 129
Fox, C. 6
France 34, 44–5
   eco-logistics *see* eco-logistics
      improvement in France
   ECODAS case study *see*
      entrepreneur profile and
      sustainable innovation strategy,
      ECODAS case study, France
   EDF Group, Web 2.0 website (Ma
      Maison Bleu Ciel d'EDF)
      innovation 272–4
Frankel, C. 264
Fuller, D. 264

Gani, A. 126
Garand, D. 188, 189

Garg, A. 93
Garnett, N. 204
Gary, I. 152
Geels, F. 31, 48
Geindre, S. 191, 199
Gendron, Corinne 51–73
General Electric (GE), Ecomagination
   programme 12
Germany, logistics management *see*
   logistics management, integrating
   sustainability and technology
   innovation, Germany
Ghalib, K. 155
Gillin, M. 175
global climate change, sustainable
   innovation responses to 3–20
   biological impact of climate change
      6
   business model changes, need for 18
   circuit board cleaning 13
   data centre energy consumption
      13–14
   eco-innovation (EI) 16
   ecosystem resources as sustainability
      challenge 10–11
   electric mobility 250–52
   employee involvement 17–18
   energy use as sustainability challenge
      10
   greenhouse gas emissions 4–5
   habitat fragmentation 6, 7
   hotels and ecological footprints 15
   ink-efficient fonts 13
   multinational corporations (MNCs)
      11–12
   oceans, climate change effects on 6
   organizational systems, changes to
      18
   physical evidence of global climate
      changes 5–6
   pollution as sustainability challenge
      10
   population as sustainability
      challenge 10
   precursors to unleashing innovation
      17–18
   recycling and industrial ecosystems
      14
   social benefits 16
   supply-chain sustainability 14–15

sustainability challenges 10–11
sustainability definition 9
sustainability as response to climate
    change 6–9
sustainable innovation overview
    11–16
transportation sector 212–13
trucking industry, idling rest stops
    13
waste management practices 14, 15
see also environmental concerns;
    greenhouse gas (GHG)
    emissions
globalization effects
logistics management, Germany
    240–41
markets and co-venturing in
    developing countries 146–7
Goberville, E. 6
Godard, O. 64
Goodland, R. 55
Göpfert, I. 240
Grameen Bank 139, 160–61, 177
green economy
construction technology in building
    sector 207
definition (UNEP) 53
eco-innovation *see* eco-innovation
    and green industry
    transformation in OECD
    countries
green and clean technology
    implementation, developing
    countries 152
green architecture, Ipsos analysis 271
political dimension, lack of 56
transition to 51–7
see also economic effects;
    environmental concerns
greenhouse gas (GHG) emissions 4–5,
    23–4, 33, 35, 38, 44–5
logistics management 246–7, 248,
    249, 250–52, 254–6
see also global climate change,
    sustainable innovation
    responses to; pollution
Group of 20 (G-20) 106–7
Guillaume, J.-P. 235
Guimaraes, Renato 219–38
Gunasekaran, A. 222

Haanaes, K. 3
Hackett, M. 142
Hairong, Y. 159
Halldorsson, A. 252
Halme, M. 42
Hambrick, D. 92
Hamet, J. 213
Hammer, M. 220
Hammond, A. 146
Harland, C. 221
Hart, S. 142, 144–5, 146, 190, 263, 264
Hazelton, P. 208
Hellström, T. 29, 30
Hemmati, M. 155
Hermalin, B. 92
Herminghaus, H. 255
Hermoso de Mendoza, A. 158
Herremans, I. 200
Hill, R. 205
Hillbrand, T. 240
Hilton Hotels, LightStay measurement
    system 15
Hlady Rispal, M. 193
Ho, M. 176
Hoa, T. 211
Hockerts, K. 145
Holt, D. 142, 160
Honig, B. 175
Hopkins, R. 158
Horsley, A. 206
Hossain, F. 155
Houé, Thierry 219–38
Hout, T. 229
Hoxha, A. 156
Hwang, B. 204

IBM 13–14, 32, 33, 35
India
Arvind Eye Hospitals 132
corporate governance *see* corporate
    governance norms and Indian
    corporate enterprises
cultural idiosyncrasies and business
    decision making 131–3
early iron technology, effects of 122
HMT rice variety, development of
    125–6
Indian Partnership Act 75
Lijjat Papad cooperative *see*
    under entrepreneurship

development at small scale as key to sustainable economic development
National Innovation Foundation 126–7
online marketing 127
open source initiative 127
outsourcing, benefits 130–31
information and communication technologies (ICT)
eco-innovation and green industry transformation 33, 38–9, 45–6
eco-social business in developing countries 142, 155
Internet access and entrepreneurship development at small scale 177
logistics management 244–5
online marketing, India 127
innovation
dropout problem, appropriate technology movement 129
eco-innovation *see* eco-innovation and green industry transformation in OECD countries
and environment–economy debate 62–7
invention increase, appropriate technology movement 121, 124, 125–8
marketing *see* marketing innovation
social challenge *see* social challenge of sustainable development through innovation
Inoue, C. 153
International Monetary Fund (IMF) 106
International Union for Conservation of Nature (IUCN) 154
Ipsos analysis and market innovation *see* marketing innovation, Ipsos analysis
iron and steel industry 32, 33, 35
ISO 26000 standard of social responsibility and sustainable development 57–60, 61–2
Israel 45–6
Italy 86–7, 156
Ivanko, J. 141

Jackson, E. 92
Jakubowski, M. 127
Jaouen, A. 186
Japan 25, 32, 36, 86
Jasper, A. 257
Jeanrenaud, S. 154
Jensen, M. 93
Jeppesen, S. 147
Johnson, S. 171
Jordan, D. 211
Julien, P. 187
Jump, A. 6, 7

Kakoty, Sanjeeb 118–35
Kalina, J. 157
Kapp, W. 66
Kar, Rabi Narayan 74–99
Karl, T. 152
Karnani, A. 147
Katerere, Y. 158
Kelley, T. 129
Kemp, R. 26
Kennedy, T. 143, 150
Keoleian, G. 264
Khasreen, M. 208
Khobragade, D. 125–6
Kibert, C. 205–6
Kiehl, J. 247
Kille, C. 239, 240
Kirkby, C. 157
Kirzner, I. 172
Klaus, P. 239, 240, 241, 245, 246, 248, 252
Klein, A. 92
Klein, P. 213
Klein, R. 213
Klumpp, Matthias 239–61
Knoeber, C. 92
Kolk, A. 264
Korkmaz, S. 209
Korten, D. 128
Kothari, L. 75
Krieger, W. 241, 245, 246, 248, 252
Krishnan, M. 263

La Pira, F. 175
Lachance, R. 187
Lambert, D. 222
Latour, B. 191
Laufer, J. 191

Lawrence, J. 92
Le Bas, C. 200
Le Boulch, Gaël 262–76
LEED (Leadership in Energy and
    Environmental Design) 14–15
Leitner, A. 152
Lemoine, O. 227
Lessem, R. 158
Lipton, M. 93
Littlewood, D. 142, 160
Littman, J. 129
logistics, eco-logistics *see* eco-logistics
    improvement in France
logistics management, integrating
    sustainability and technology
    innovation, Germany 239–61
  business cycles, effects on 256
  ecological perspective 246–8
  economic perspective 241–6
  global supply chains 241, 245–6
  globalization effects 240–41
  greenhouse gas emissions 246–7,
      248, 249, 250–52, 254–6
  greenhouse gas emissions, power
      plants comparison 255
  information technology, importance
      of 244–5
  integrated development perspective
      257
  logistics market 239–41
  oil prices and dependency 250
  outsourcing 245–6
  respirable dust emissions, reduction
      in 248
  risk management and secure
      handling of goods transport
      247
  service levels 244–5
  sustainable solutions, improvement
      of 241–2, 248, 252
  vendor-managed inventory 246
  warehouse centralization concepts
      242–4, 245–6, 256
  waste management 247
logistics management, integrating
    sustainability and technology
    innovation, Germany, electric
    mobility 249–56
  battery power duration 253–4
  and climate change 250–52

and demographic change 252
environmental balance 254–6
and resource protection 253
urbanization factors 252–3
Loorbach, D. 48
Lorsch, J. 93
Louche, C. 144
Lozano, J. 191
Lydenberg, S. 144

McCormick, J. 154
McDonough, W. 145
Machiba, Tomoo 21–50
McKie, D. 191
McKinnon, A. 220, 228
McMullen.J. 140, 150
McWilliams, A. 186
Maibach, M. 231, 232
Mair, J. 172, 173, 174, 175, 176, 177,
    178
Mak, Y. 93
Makita, R. 143
marketing innovation 262–76
  customer relationship management
      (CRM) 263–4, 272–4
  Designfor-X process 263–4
  EDF Group, Web 2.0 website (Ma
      Maison Bleu Ciel d'EDF)
      innovation 272–4
  environmental management systems
      (EMS) quality standards (ISO
      14000) 265
  environmental new product
      development theory (ENPD)
      and eco-performance 263–4,
      265
  recycling 264
  see also corporate governance
      norms and Indian corporate
      enterprises; multinationals,
      codes of conduct and other
      multilateral control systems for
marketing innovation, Ipsos analysis
    265–72
  back to sources trend 266–7
  carbon offset schemes 270–71
  customer transparency 269
  ethical concerns trend 269–70
  financial ethics 269–70
  green architecture 271

guilt absolution trend 270–71
holistic green trend 271–2
local materials, importance of use
    of 267
natural concept, expansion of 267
pollution reduction 268
raw materials, use of 266–7
revival of older techniques 269
voluntary reduction trend 267–9
voluntary simplicity 268
waste management and recycling 268
Marti, I. 172, 173, 174, 175, 176, 177,
    178
Martin, M. 46
Martin, R. 174
Martinet, A. 190, 194, 262–3
Mashelkar, R. 132
Masi, A. 149, 156
Massaro, Maurizio 139–67
Masurel, E. 141
Mathieu, A. 186, 189, 190, 194, 195
Maxwell, D. 222
Meadows, D. 63
Meenakshisundaram, R. 46
Mehta, R. 207
Meiarashi, M. 213
Melo, M. 229
Menon, A. and A. 141, 264
Mentzer, J. 220
mergers and acquisitions 75, 90, 91
    see also multinational corporations
        (MNCs)
Meunier, C. 228
Mexico 158
Michelin, eco-innovation 32
Miles, M. 264
Miles, R. 191, 194, 200, 229
Milstein, M. 142
Mohamed-Katerere, J. 158
Mohammed, O. 177
Molteni, M. 156
Morin, X. 6
Morris, M. 172
Mort, G. 174
Mortensen, O. 227
Mowforth, M. 157
Muchner, C. 241
multinational corporations (MNCs)
    global climate change, sustainable
        innovation responses to 11–12

involvement in eco-social business in
    developing countries 155–6
mergers and acquisitions 75, 90,
    91
multinationals, codes of conduct and
    other multilateral control systems
    for 100–117
collective approach, need for 102
corporate misconduct, recent 101–2,
    112–15
Group of 20 (G-20) involvement
    106–7
history of 101
International Monetary Fund (IMF)
    involvement and financial
    transparency 106
OECD involvement 101, 107–8
UN Centre on Transnational
    Corporations (UNCTC) 103,
    111
UN Centre on Transnational
    Corporations (UNCTC), code
    of conduct draft 105, 116–17
UN Conference on Trade and
    Development (UNCTAD),
    developing countries and FDI
    105–6
UN Global Compact 103–4
UN involvement 103–6
World Trade Organization (WTO)
    involvement 107
    see also corporate governance
        norms and Indian corporate
        enterprises; marketing
        innovation
Munilla, L. 264
Munt, I. 157
Mutagwaba, B. 143

Nel, E. 157
Nelson, R. 3, 200
Netherlands 68–9
Neumayer, E. 145
Ngai, E. 222
NGO involvement 69–71, 156
Nicaragua 148
Nicholls, A. 175
Nicolau, J. 130
Nidumolu, R. 3, 67
Nigeria 156–7

Nilekani, N. 133
Nordic Council of Ministers 39, 40
Norton, M. 270
Noya, A. 149

Obot, A. 155
Obot, I. 155
Ochoa, L. 206
OECD involvement
    eco-innovation *see* eco-innovation
        and green industry
        transformation in OECD
        countries
    multinationals, codes of conduct
        and other multilateral control
        systems for 101, 107–8
Offner, J.-M. 227
oil prices 225, 250
Organisation for Economic Co-
    operation and Development
    (OECD)
    Green Growth Strategy 21–2, 47
    Oslo Manual of innovation (OECD/
        Eurostat Oslo Manual) 25–6
Orsato, R. 219
Osberg, S. 174
Osei-Hwedie, K. 160
Oshisanya, K. and T. 152
Ostertag, M. 257
Osterwalder, A. 40
Oudghiri, Remy 262–76
outsourcing 130–31, 245–6

Paché, G. 229
Palazzo, G. 60
Palmer, K. 66
Pape, J. 93
Patris, C. 189
Pauli, G. 145
Pearce, A. 208–9
Pearce, D. 55, 144
Pearson, P. 26
Peattie, K. 263, 264
Pennisi, E. 212
Penuelas, J. 6, 7
Perrons, D. 152
Peter, R. 177
Pfohl, H. 240–41
Phillips, R. 213
Pickrell, S. 204

Piebalgs, A. 152
Pigou, A. 66
Pinchot, G. 141–2
Pinstrup-Andersen, P. 149
policy considerations
    eco-innovation and green industry
        transformation in OECD
        countries 44–6
    policy entrepreneurship and eco-
        social business in developing
        countries 149
    regulatory framework, Indian
        corporate enterprises 79–80, 87,
        94–5
    transportation sector 214
pollution concerns
    air-pollution-absorbing concrete and
        bricks 213
    global climate change, sustainable
        innovation responses to 10
    pollution reduction, Ipsos analysis
        268
    social challenge of sustainable
        development through
        innovation 64, 65
    see also greenhouse gas (GHG)
        emissions
Ponnekanti, J. 157
Porter, M. 24, 64–6, 67, 264
Postman, N. 118, 119, 120, 121
poverty reduction
    bottom/base of the pyramid (BOP)
        strategies 146–8, 150, 171, 175,
        176–7
    and business entrepreneurship 150,
        153, 155–6
    and foreign investment 147–8
    India 91–2
    and social challenge of sustainable
        development through
        innovation 54, 57
Powell, W. 190
Prahalad, C. 146, 148, 176, 182, 263
Prinsen, G. 159
production system
    closed-loop 28–9
    efficiency measurement and profit
        motive 120–21
    inland waterway transport 231–3
    product-service systems (PSSs) and

green industry transformation
39, 40, 42
profit-making 120–21, 175, 176, 181
Prowell, B. 212
Pujari, D. 263

Qian, Q. 207

Randjelovic, J. 141
Rasmussen, R. 213
Raynor, M. 128–30
recycling 14, 145–6, 151–2, 211–12,
264, 268
see also environmental concerns;
waste management
regulatory framework *see* policy
considerations
Reuther, J. 253
Reveret, J. 55
Reynaud, E. 186, 190, 194, 262–3
Rhodes, P. 129
Richardson, B. 211
Ricupero, R. 101
risk management
eco-innovation and green industry
transformation in OECD
countries 31
eco-logistics risks, inland waterway
transport 231
fundamental safety risk levels,
building sector 209
logistics management and secure
handling of goods transport
247
Rist, G. 57
Robinson, J. 172
Rondinelli, D. 13
Rosenberg, N. 128
Rostow, W. 53
Roy, M. 77
Ruef, M. 152
Rugraff, E. 147
Ruli, G. 156
Russel, S. 70

Sagafi-nejad, Tagi 100–117
Salles, J. 64
Salomon, J. 64
Samoilovich, Y. 213
Sampath, K. 80

Santos, F. 173, 175
Sanya, T. 151–2, 158
Sassen, S. 253
Sautman, B. 159
Savy, M. 222, 227
Schaeffer, L. 174
Schaltegger, S. 141
Schaper, M. 141, 142
Scherer, A. 60
Scherhorn, G. 144, 145
Schieffer, A. 158
Schmidheiny. S. 220
Schmidt, J. 223
Schmitt, Christophe 186–203
Schoenberger-Orgad, M. 191
Schönwiese, C.-D. 250, 251
Schramm, C. 176
Schumacher, E. 121–2, 123, 124
Schumpeter, J. 168, 171, 188
Schützenmeister, F. 252
Scrase, I. 30, 31
Sehgal, K. 147
Seyfang, G. 70, 71
Shankar, B. 46
Sharma, R. 122
Sharman, M. 264
Sharp, eco-innovation 32
Sheshabalaya, A. 130, 131, 133
Shibata, H. 213
Shrivastava, Paul 3–20, 262, 264
Siemens VAI, eco-innovation 32
Signore, J. 213
Silitshena, R. 160
Singh, A. 209
SMEs *see* entrepreneurship
development at small scale as
key to sustainable economic
development
Smith, A. 70, 71
Smith, K. 31
Snow, C. 191, 194, 200
social business *see* eco-social business
in developing countries
social challenge of sustainable
development through innovation
51–73
competitiveness and regulatory
standards 64–7
ecologist movement and
technological innovation 63–4

economic crisis effects 54
economic growth, challenge of 53–6, 60–61, 63
environmental externalities factor 66
environmental protection in developing countries, challenge of 52–3
grass-roots innovations 69–71
green economy definition (UNEP) 53
green economy, political dimension, lack of 56
green economy, transition to 51–7
innovation and environment–economy debate 62–7
innovation and 'society pull' (transition management) 68–9
innovation for sustainable development 67–71
innovation subsidies 68
ISO 26000 standard of social responsibility and sustainable development 57–60, 61–2
NGO involvement 69–71
pollution concerns 64, 65
Porter Hypothesis on regulation and competitiveness 64–6, 67
poverty eradication 54, 57
social dimension of sustainable development 56–7, 69
social responsibility and sustainable development 57–62
social responsibility concept, history of 60–61
socio-technical regime proposal 69–70
sustainable development, confusion in understanding of 51–3, 55, 57–8
social entrepreneurship *see* entrepreneurship development at small scale as key to sustainable economic development, social entrepreneurship
social impact
appropriate technology movement 125–6
challenge of sustainable development through innovation 55–6

eco-innovation and green industry transformation in OECD countries 26, 31
Soparnot, R. 186, 188, 190, 194, 195
Sparkes, R. 144
Spence, L. 191
Spring, A. 155
Stake, R. 193
Stalk, G. 229
Stan, C. 252
Stapledon, G. 92
Steinert, J. 207–8
Stevens, E. 186, 188
Steward, F. 31, 129
Stiglitz, J. 110
Straube, F. 240–41
Subway restaurant chain 14–15
Sullivan, R. 219
supply chains 14–15, 219–23, 241, 245–6
Swaney, J. 64
Sweden 46
Swiaczny, F. 253
Switzerland 156

Taylor, F. 120
technology, appropriate *see* appropriate technology movement
technology innovation and sustainability, logistics management *see* logistics management, integrating sustainability and technology innovation, Germany
Tecnosol 148
Ten Hompel, M. 244, 246
Thomas, C. 11
Tischner, U. 44
Toutanji, H. 211
Toyota, eco-innovation 32
transnational companies (TNCs) *see* multinational corporations (MNCs)
transport industry 32, 33, 35, 38, 44–6
electric mobility *see* logistics management, integrating sustainability and technology innovation, Germany, electric mobility

inland waterways *see* eco-logistics improvement in France, inland waterway transport
sustainability *see* construction technology in the building and transportation sectors, benchmarking, transportation sector
trucking industry, idling rest stops 13
Tremblay, D. 188, 192
Trenberth, K. 247
Tukker, A. 44, 48

Uganda 158
Ullrich, R. 247
ULSAB-AVC, eco-innovation 32
Ummenhofer, M. 230
United Arab Emirates, sustainable technology use in commercial and residential projects case study 214–15
United Nations
Centre on Transnational Corporations (UNCTC) 103, 105, 111, 116–17
Conference on Environment and Development (UNCED) 8–9, 153–4
Conference on Sustainable Development, Agenda 21 154–5, 205
Conference on Trade and Development (UNCTAD), developing countries and FDI 105–6
Framework Convention on Climate Change (UNFCCC), Kyoto Protocol 9, 249
Global Compact 103–4
Human Settlements Programme UN-HABITAT 8
Montreal Protocol on Substances that Deplete the Ozone Layer 7, 8
multinationals, codes of conduct and other multilateral control systems for 103–6
Stockholm Declaration on the Human Environment 52

United Nations Environment Programme (UNEP) 8
Global Green New Deal (GGND) 54
green economy definition 53
Green Economy Report 21, 53–4, 56, 57, 66–7
Intergovernmental Panel on Climate Change (IPCC) 250–51
World Conservation Strategy (WCS) 154
United Parcel Service (UPS), idling savings 12–13
US
construction industry 205, 206
Dodd–Frank Wall Street Reform and Consumer Protection Act 102
eco-innovation 24
Foreign Corrupt Practices Act (FCPA) 101
Green Building Council 14
residential energy consumption 206
Sarbanes–Oxley Act 101
Securities and Exchange Commission (SEC) 102
transportation construction and use of recycled materials 212
Ultra-Light Steel Auto Body (ULSAB) initiative 35

Van de Ven, A. 188
Van der Horst, R. 222
Van der Linde, C. 24, 64–6, 264
Vélib', bicycle sharing system 32, 34
Ventzke, R. 246
Verkaar, H. 213
Verma, J. 93
Verstraete, T. 187
Vinacke, H. 123
Vogelpohl, A. 253
Vollenbroek, F. 67–9
Von Weizsäcker, E. 145, 151

Wacheux, F. 193
Wad, P. 147
Wagenhofer, P. 255
Walubita, L. 211
Wang, J. 145
Ward, A. 213

waste management
  building sector 208
  global climate change, sustainable
    innovation responses to 14, 15
  logistics management 247
  and recycling, Ipsos analysis 268
  see also recycling
Wayson, R. 213
Weirner, J. 93
Weisbach, M. 92
Weller, G. 6
Welsh, C. 200
Wernerfelt, B. 186
Wheeler, W. 264
White, L. 123, 124
Williams, A. 159
Williams, O. 109
Williams, R. 70
Winter, S. 3, 200
Wong, V. 264
Wood, D. 60
World Business Council for Sustainable
    Development (WBCSD) 9, 39

World Health Organization (WHO)
  8
World Meteorological Organization
  (WMO) 5, 250–51
World Trade Organization (WTO)
  107
World Wildlife Fund (WWF) 154
Wüstenhagen, R. 152

Xerox, eco-innovation 32, 35

Yermack, D. 92, 93
Yigitcanlar, T. 209
Yin, R. 193
Yokogawa Electric 32, 36
Young, A. 147
Yuanto, K. 93
Yunus, M. 139, 148, 160–61, 177

Zahra, S. 150, 171
Zelewski, Stephan 239–61
Ziegler, M. 252
Zott, C. 41